Jan., 2014

STILLWATER
PUBLIC LIBRARY

Stillwater, Minnesota 55082

THE FALLING SKY

Also by Ted Nield

Supercontinent: Ten Billion Years in the Life of Our Planet

THE
FALLING SKY

The science and history of meteorites
and why we should learn to love them

TED NIELD

LYONS PRESS
Guilford, Connecticut
An imprint of Globe Pequot Press

Copyright © 2011 by Ted Nield

First published in the UK in 2011 as *Incoming!* by Granta Books

First published in the USA in 2012 by Lyons Press, an imprint of Globe Pequot Press

The list of illustrations on pp. 275–77 constitutes an extension of this copyright page.

Designed by M Rules

Library of Congress Cataloging-in-Publication Data is available on file.

ISBN 978-0-7627-7828-7

Printed in the United States of America

10 9 8 7 6 5 4 3 2 1

Pour voir les choses, il faut les croire possible.
(In order to see things, it is necessary to believe them
possible.)

MARCEL ALEXANDRE BERTRAND, GEOLOGIST

Some things have to be believed to be seen.

RALPH HODGSON, POET

CONTENTS

INTRODUCTION   IX

A NOTE ON TERMINOLOGY   XIV

PART ONE—DREAMTIMES

1   A SERIES OF UNFORTUNATE EVENTS   3

2   STARRY MESSENGERS   29

3   THE FALLING SKY   61

PART TWO—DEMONS

4   TARGET—EARTH   95

5   DEEP IMPACT   122

6   DESTROYERS OF WORLDS   140

PART THREE—DELIVERANCE

7   LIFE IS EVERYWHERE   177

8   RAIN FROM HEAVEN   202

9   A KICK IN THE GENES   221

10 UNREST CURE   245

EPILOGUE   258

ACKNOWLEDGMENTS   267

FURTHER READING   270

LIST OF ILLUSTRATIONS   275

INDEX   278

INTRODUCTION

I deas flash across human minds all the time. However, only some of those flashes of inspiration become historically significant for everyone, rather than for a mere individual. As the story of meteorites unfolds, it will show how the significance of a scientific idea, just like that of any particular meteorite, depends crucially upon whose head it falls and when. People have been trying to understand shooting stars and meteorites (shooting stars that survive atmospheric entry and reach the ground) for as long as they have stared into space; and our continuing attempts to derive meaning from these intriguing objects is what this book is all about. Long before natural science as we understand it existed, people looked at nature and asked *why*—usually in an attempt to divine the future. Science turns this instinct on its head. Science does not care why; it asks *how,* and the answers to that much more fruitful question deliver information about the past.

Scientists—and especially geologists—look at things as they are today and, by asking how they came to be that way, see process. Charles Darwin was first and foremost a geologist; when he looked at living species, instead of seeing separate entities, each specially created in its perfection and packaged up in a box, he saw a messy historical process—organic evolution—that stretched behind them deep into time, connecting all living things with their common ancestors. In a similar way, meteorites connect us with our deepest origins, as one part of the planet that, with more than a little help from the cosmos's starry messengers, gave us life.

When you look into space, you gaze upon the abyss of time—and time is something that I and my geological companions felt we understood. We knew, for example, that the limestone rocks we were mapping were about 55 million years old. We knew they had formed in a warm shallow sea 35 million years before the jagged mountains (that now reared up behind us like the walls of Mordor) had even begun to rise. As soon as they did, erosion began to wear them down, and the thick layers of desert-varnished boulders that now buried the uplifted limestones had swept down, during a billion torrential storms, right up until the present day. As scientists do, we looked at our rocks and saw process. This set the limits of our local geological time frame for the latest 50 million of the 4.567 billion years that have passed since the Earth, the other planets, and the meteorites formed from the ashes of dead stars.

And yet that night sky above us was older than everything, taking us almost as far back as the origin of the universe, 13.5 billion years ago. Such were the distances into which we gazed that much of the starlight entering our eyes had begun its journey long before all of that Earth history had happened; long, indeed, before even our Solar System—the Sun and the orbiting planets—had begun to condense. Some of the light had been born in stars that (though the news of it has yet to reach us) died cataclysmically billions of years ago, blowing new, heavy elements across the void so that bright young suns, some with orbiting planets, could arise from their ashes. The story of meteorites begins then—at the time, within the longer history of the cosmos as a whole, that our Sun and its circling worlds came into being.

Since that far-off time, a few tens of thousands of tons of meteoritic material have been dropping to Earth each day—time capsules from that otherwise untouchable moment 4.567 billion years ago. As material objects, they are older than anything on Earth—the oldest things, in fact, that you can possibly hold, so continually changed and recycled is all the rocky stuff beneath our feet on our geologically active planet. It is hardly surprising that throughout human history meteorites—with their peculiar appearance and spectacular arrival—have become objects of myth. Our time differs from those that have gone before because we are now two centuries into an age where science dominates our understanding of nature. The stories that we build around meteorites today, principally as agents of doom and death, are scientific, concerning the history of our planet and its life. But for all the added rigor of science, these tales remain constructions of the human imagination, and therefore derive (in part at least) from our collective history and worldview and from the social and political tenor of our age.

Since 1980, with the discovery that the mass extinctions of the end-Cretaceous were accompanied by a massive impact, we have become aware that meteorites not only built our world, but that larger ones retain the power to change its destiny as agents of almost unimaginable destruction. This revolution in our understanding came as a final legacy of World War II as we entered the paranoia of the nuclear age with its ICBMs visiting "megadeath" upon us from above. Today, in a different world, scientists are discovering that meteorites have also brought life to Earth—though not in a way that anyone ever suspected. Many have pursued the idea that life pervades the cosmos and that its manifestation on Earth grew from germs sneezed upon a receptive world like cosmic flu. Now, instead, scientists are beginning to think that disturbance of Earth's ecosystems by large but sub-lethal meteorite impacts might have afforded new evolutionary opportunities to home-grown life—opportunities that led to greater diversity and complexity and ultimately to us.

How are these two ideas about meteorites, one of death and another of life, reconcilable? The deceptive simplicities of our politically and culturally divided past have given way to an era when we, thinking for the first time as a species, are realizing that we must act together to survive. Meanwhile, apparent scientific certainties, built on the fault-lines that ran between the traditional disciplines, are being abandoned as we come to a more holistic understanding of how the Earth system operates and has evolved through geological time. Earth sciences are today embracing the idea that no significant event in the past can ever truly be said to have had a single cause—any more than a single cause will always produce the same effect on a complex and evolving system like the living Earth.

This story of this interaction of Earth-life and its cosmic environment can make sense only if we allow that within a complex whole like our ever-changing planet, a particular cause may produce dramatically different effects—say, on life 460 million years ago—than it might have done just under 400 million years later, when the dinosaurs vanished. This concept is not difficult to grasp; it is easy to see how a particular event might produce dramatically different consequences in one culture from another or mean something different to two individuals with contrasting views about the world. The music of the spheres is but a single, repeating theme. The history of our planet is a symphony, and, in the story of how we have come to appreciate the complex counterpoint between the two, we hear the greatest instrument of all: the human voice.

A NOTE ON TERMINOLOGY

The mass extinction that carried off the dinosaurs, along with much else, 65 million years ago, features prominently in this book. The event has been used as a marker for the end of the geological period called the Cretaceous, which began 145 million years ago in succession to the Jurassic Period.

Where one period ends, another must begin; so 65 million years also marks the base of the next period, which is the Palaeocene. Clearly, finding a shorthand way of referring to this important boundary was desirable, and it became customary to call it the K-T Boundary. This may seem a little odd to English-speakers; but the K derives from *Kreide*, the German equivalent of Cretaceous, while the T comes from the term *Tertiär* (Tertiary, in English), which is a very old term for what came next.

The term K-T Boundary has become widely recognized, even among the lay public, so notwithstanding its various shortcomings I use it throughout this book—unashamedly eschewing recent pedantic attempts to replace it with "K-P," anticipating the International Commission on Stratigraphy's awaited decision on whether to consign the Tertiary to history.

PART ONE

DREAMTIMES

1

A SERIES OF UNFORTUNATE EVENTS

Alien they seemed to be:
No mortal eye could see
The intimate welding of their later history.

THOMAS HARDY, "THE CONVERGENCE OF THE TWAIN"

The historical meaning of any event that befalls nature derives principally from the context in which it happens, just as for events in human history, and just as our view of what meteorites mean has changed through history, according to the milieu into which the starry messengers have so unexpectedly dropped. There is no knowing how societies or individuals will react to a great event until it happens.

Just after 4:00 p.m. on Christmas Eve 1965, stars appeared over Leicestershire, in the heart of the English Midlands. Initial reports were confused—especially as the reported fireballs and sonic booms were, at first, assumed to refer to a single celestial light like the one in the Bible. Furthermore, as a wise man named Jack Meadows later wrote: "Strange sights and sounds are not that uncommon on Christmas Eve."

Meadows, an astronomer with a special interest in what he has called "space garbage," has worked for most of his career in the University of Leicester's Department of Astronomy. Like many in this field, he has the honor of having an asteroid named for him—4600 Meadows. On hearing the reports,

Meadows guessed that a meteorite fall had occurred more or less in his back yard, and he raced to the scene—Barwell, an industrial village just outside Leicester. From the many eye-witness accounts he collected that freezing January, Meadows and a colleague worked out that there had been three major fireballs descending at about 20 degrees to the horizontal, each created by a fragment of a single meteorite.

As it first hit the top of the atmosphere, it would have been traveling at about 12.5 miles per second. The initial shock of slamming into our atmosphere began the process of disintegration. Air and water seem gentle to us, wallowing as we do through the molecular soup that everywhere covers the Earth. When walking, swimming, or even parachuting, our atmosphere flows around us more or less easily. But suddenly hitting fluids at very high speed is quite different, as anyone knows who has done a belly-flop from a diving board. An aircraft crashing into the sea might as well be hitting concrete, and the same goes for meteorites as they collide with even the thin upper atmosphere at cosmic velocity. A meteorite has to be immensely strong to withstand these stresses without disintegrating—a fact which, incidentally, affects the abundances of the different types of meteorite that survive the fall intact.

The Barwell meteorite was made of rocky material and so broke up rather easily as it tore through the air. The surfaces of its fragments became incandescent, melting away from their leading edges, creating thin crusts of fused glassy rock that became sculpted into thumbprint-like hollows. At the trailing edges, molten rock flew off in a spray of silicate droplets, which solidified immediately, and would have hung in the air for hours or even days before finally dropping unheeded to

Earth—joining a planet at long last, after wandering the Solar System for 4.567 billion years.

Of the fireballs, the easternmost fizzled out somewhere between Rugby and Leicester. The other split into two to the southwest of Warwick. One of these ended in Barwell; the other farther north, beyond Loughborough. But only the Barwell fireball led to a cache of fallen space rocks.

The points where these three fireballs appeared to the observers to go out, known as extinction points, when the fragments finally lost their cosmic speed and became mere falling stones, would have occurred at about the cruising height of a jet aircraft (between 5.5 and 6 miles high). The fate of incoming cosmic debris is largely a function of size. Dust-sized particles have so little momentum that they may simply float undamaged into the upper atmosphere. Grit and sand-sized particles burn up high above and form the shooting stars. Larger objects retain more of their original speed deeper into the atmosphere. Some burn up completely. Others, like Barwell, burn partially and then drop to Earth under gravity.

Very large objects, like the one that saw off the dinosaurs, preserve some or even all of their cosmic velocity all the way to the ground, dissipating immense kinetic energy as they hit. However, such impactors are mercifully rare by human standards, and it is the fate of most rocky meteorites (by far the most common sort) to succumb to the forces tearing them apart and to litter the Earth with their wreckage.

Barwell's final disintegration led to a shower of smaller stones hitting a roughly circular area covering several square miles to the south of the village. One piece broke obligingly through a factory roof and several floors below, enabling Meadows to work out that the angle of descent steepened to nearer 80 degrees as cosmic velocity ebbed away and gave way to gravity. Altogether, the pieces recovered—some of which fitted together like a 3D jigsaw puzzle—weighed 45 pounds. It was the largest meteorite fall ever recorded in the United Kingdom.

Meadows quickly interviewed local residents about their experiences. The first fall was witnessed by Arthur Crow. Returning from his work at Harvey & Co.'s shoe factory, he heard a number of stones landing in quick succession. "Half a dozen pieces came down and at first it seemed like rockets. The piece which fell outside the house could have knocked anyone's head off or smashed the roof in," he told a reporter from the *Leicester Mercury*. "I crouched against a wall and then saw powder rising from the road."

One fragment, shattering on impact, broke the front window where Mr. Joseph Grewcock lived. "I picked up a piece of rock and dropped it again quickly because it was still hot. It was very heavy and black on one side . . . from burning," he told the newspaper. The window-smashing fragment itself was eventually found lying in the pot of a houseplant.

Jack Meadows recalled that "One man had the bonnet of his parked car hit . . . and had initially been sure it had been thrown by a couple of boys up the road." This unfortunate gentleman was Percy England. "It did not cheer him up to find he was wrong," Meadows wrote. "Vandalism came under his insurance cover, but a meteorite was reckoned an act of God." Subsequent accounts suggested that Mr. England later went and demanded payment from the local vicar, but this is probably embroidery.

In modern times, motor cars feature quite a lot in tales of meteorite falls—which is not really surprising, because they are fairly large and spend nearly all their time outdoors. Some owners have found good fortune in their celestial visitors; but sadly Mr. England did not. So convinced had he been by his vandalism theory that he disposed of the "piece of concrete" (which weighed about 6.5 pounds) on waste land. Later, he heard that ample rewards were being paid by London's Natural History Museum for fragments of the meteorite—five shillings an ounce for anything weighing less than two ounces (about one new penny per gram) rising to seven and sixpence per ounce for anything up to one pound weight, and a princely ten shillings per ounce for anything bigger. All was perhaps not lost. But alas, when Mr. England returned to the waste land, he was unable to find the now precious object. In the meteorite gold rush, as the *Leicester Mercury* called it, somebody had beaten him to it. And so it has been throughout history that witnesses to meteorite falls have judged the meaning of the event according to the context of their own lives, thoughts, and attitudes. And so it is, on a small and human scale, that events can always be said to take their meaning from the context in which they occur, whether the event in question affects individuals, societies or even planets.

The museum's top payout was £140 for a 17-pound fragment. Harold Platts, who was thirty-seven at the time of the fall, found his 5-pound fragment "digging around the sprouts" on his allotment. He took the piece home, weighed it, and placed it carefully on his piano. At the going rate decreed by the redoubtable Dr. G. F. Claringbull, keeper of minerals at the Natural History Museum, Harold got £39 10s—which was enough to take him, Mrs. Platts, and their two children on a weeklong vacation in Great Yarmouth.

The recovered fragments were soon subjected to radiometric dating and shown to share with most other meteorites the maximum possible age that any object in our Solar System large enough to hold in your hand can show—about 4.5 billion years, the date when most solid matter first condensed from the solar nebula. But meteorites, like very old buildings or pieces of furniture, have been changed, added to, divided, and reassembled at various stages in their long lives. So although most meteorites tell us when our Solar System began to take its present form, the long period of their existence in between—including the entire geological history of our planet and of its rocky neighbors, Mercury, Venus, and Mars—was rarely without incident.

Most meteorites can be shown to have another "age," which dates the moment when they were liberated from their parent body (usually, though not exclusively, an asteroid). Which is why, to tell our next impact story and fully explain how Michelle Knapp got a new car, we must go back in time to a violent event that occurred 50 million years ago, somewhere in the Asteroid Belt—that broad planetless zone between Mars (outermost of the rocky planets) and Jupiter, the first and largest of the gas giants.

To get a feel for what 50 million years ago means, our world was then in what geologists call the early Eocene Period. Modern mammals were just starting to appear ("Eocene" means "dawn of the recent')—a faunal switch coming about as the Earth underwent an intense period of global warming called the Palaeocene–Eocene Thermal Maximum (PETM). Although Earth was beginning its last convergence with the familiar place it is today, it would be another 40 million years or so before any of your, my, or Michelle Knapp's more direct ancestors climbed in its trees.

In the River Thames, crocodiles and hippos wallowed, and rhinoceroses roamed where as yet un-evolved naked apes would one day build a town called Barwell. As yet innocent of us, this earthly paradise was equally unaware that a single collision, over 120 million miles away among the asteroids, would one day lead to the destruction of a red Chevrolet Malibu on Wells Street in Peekskill, New York.

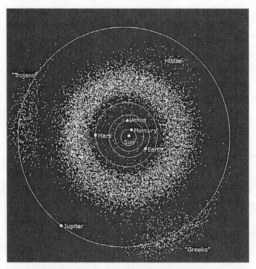

The Asteroid Belt lies between the orbits of Mars and Jupiter. This diagram does not show the Kirkwood Gaps.

That collision broke a fragment weighing perhaps 20 tons from the Peekskill meteorite's parent asteroid and sent it off along a new orbit. Different from those followed by the other asteroids in the Belt (all of which orbit the Sun in the same direction and plane as the planets), this new orbit was directed in toward the Sun, crossing the orbit of Earth. However, space is so vast and planets so minute by comparison, that 50 million years passed before collision finally happened, on Friday, October 9, 1992.

The asteroid fragment probably had a diameter of between one and two yards, and had passed its closest point of approach to the Sun (about 82.5 million miles) 41 days before it finally plummeted toward Earth. On this occasion, having successfully bypassed Mercury and Venus, it found itself heading for the Earth. Skimming the atmosphere about 25 miles too low, its 50 million years as an orphan drifter (not to mention its previous 4.5-billion-year history as one part of a remnant from planet formation) were about to end. The Earth's outer airs began to tug, slowing it down just enough to ensure that instead of skipping over the atmosphere like a stone across a pond, it cut in a little too steeply and began a downward hurtle from which escape was impossible.

It came in to land in the United States in the early evening along a northeasterly track. As it was "catching the Earth up," its relative speed was about 6 miles per second—a figure raised initially by gravity to just under 9 miles per second. Most of us are not used to dealing with these units, so if it sounds slow, 9 miles per second equates to about 33,554 miles per hour. For comparison, the Space Shuttle begins atmospheric re-entry at about 4.7 miles per second, while the fastest airliner in service, the Boeing 747-400, cruises at a mere 0.16 miles per second.

The Peekskill meteorite first appeared as a fireball over West Virginia at 7:48 p.m. local time, and crossed West Virginia, Maryland, Pennsylvania, and New Jersey on its way to New York. It was early Friday evening. People were out, enjoying the prospect of the weekend, watching football games, visiting malls and fast-food joints, shopping. For this reason, sixteen film and video recordings documented the passage of the fireball that gave rise to the Peekskill meteorite. It is these (video cameras with their handy time recorders and film cameras working at known numbers of frames per second) that have made it possible to determine the meteor's orbital trajectory and speed so precisely.

When an observer on the ground sees a meteor flying through the sky, he is seeing that line projected against the heavens. The heavens do not form a great crystal hemisphere like a snow-globe, but it helps to think of them that way. The observer is staring up from one part of the scene below, as the meteor streaks across. To determine its true path, at least one other observer needs to see the same event from a few miles away. Then the two observations can be combined mathematically to produce a true line of travel—which is where the two separate observations intersect.

To find out where a meteorite is likely to have landed, it is very important to have observations of its fireball's extinction point—the point at which it appears to go out. From that point, the meteorite ceases to fly and begins to fall. Defining this point in space also requires two independent observations of the event, several miles apart. Two widely separated observations of the extinction will describe two intersecting lines in the air, which we can think of as the hypotenuses of two right-angled triangles standing back-to-back. The length of their shared vertical side gives the height of the meteorite's entry into free-fall.

And so, at a height of just over 25.5 miles (or just inside the ozone layer) the Peekskill meteorite began its final disintegration. The pieces that peeled off from the main fireball track, clearly visible in movie footage, each had their own shapes, masses, and aerodynamic drag; and so began to fall at different rates. Sadly, if these survived as far as the ground, there they lie to this day. None was ever recovered.

As the fireball streaked by, video cameras in many states were turned skyward—at a football game in Anne Arundel County, Maryland, and in Olmsted Falls, Ohio; from a parking lot of a Burger King in Fairfax, Virginia; at the Westover Senior High football game in Raleigh, North Carolina; and by a sportscaster reporting for college station WICB at another high-school football game in Falconer, New York. The best footage of all, lasting 22 seconds, was taken by sportscaster J. Derr of WWCP, who was covering a game in Johnstown, Pennsylvania. In fact, it is safe to say that never in the history of meteoritics have astronomy and football come together so fruitfully.

All film and video footage shows a greenish fireball (the Barwell meteorite was described as "bluish green") with a flickering wake traveling northeast and at its brightest casting as much light as a full moon. The meteor covered somewhere between 435 and 497 miles in 40 seconds; though by the time the leading fragment hit its extinction point, it had slowed to a mere 3 miles per second.

There were also sounds. Sonic booms were heard over a wide area; but Patsy Keith and her family, sitting in their car near Altoona, Pennsylvania, were first made aware of the fireball by a crackling noise that they described as "like a sparkler." This lasted for about 10 seconds after the first main fragmentation event and is thought to have been "electrophonic." Electrophonic sound is what comes out of a loudspeaker; only loudspeakers are designed to create it. But electrophonic sound can come from any object capable of acting like a loudspeaker, making low-frequency electromagnetic radiation audible.

Thanks to the many video recordings, Peekskill is now one of only four meteorites in history for which a pre-fall orbit around the Sun has been determined. This is how scientists can say quite precisely where the initial collision event occurred 50 million years ago. Its orbit, determined by the observations, was an ellipse, like all orbits, and it follows that the point where the initial collision took place will lie at the point on that ellipse that lies farthest from the Sun—the point known as the aphelion. The orbital calculations showed that this lay just over twice the distance of the Earth from the Sun (or two "astronomical units") away from our neighborhood star, pinpointing Peekskill's origin at the inner edge of the Asteroid Belt. With each of its orbits lasting 1.8 Earth years, the Peekskill meteorite had probably crossed the Earth's orbit about 90 million times since its liberation.

Lying 41 miles north of New York City on a bend in the Hudson River, Peekskill is a moderately sized and relatively prosperous former manufacturing town established in the early eighteenth century; it later became the birthplace of Mel Gibson, Paul Reubens (Pee-Wee Herman), and Stanley Tucci. Just before eight that evening, as the high-school jocks performed their gridiron heroics, eighteen-year-old Michelle Knapp was sitting quietly at home when she recalled hearing something "like a three-car crash" in the street. She went out at once, but saw no crash. Instead, the nearside rear corner of her red 1980 Chevy Malibu sedan now had a mangled hole in it. Missing the fuel tank by inches, the meteorite had passed straight through the vehicle and embedded itself in the driveway below. Michelle told reporters at the time how she bent down to touch the object and noted that it was warm and smelled of sulfur. It weighed 27 pounds.

Michelle contacted the police, whose early theories included a bomb and a detached piece of aircraft. Just like the

unfortunate Percy England, they suspected foul play and filed
a report of criminal mischief. Local reports mention that it was
a neighbor who first uttered the word "meteorite," but the official version has it that Edmond Walker of the Lamont-Doherty
Earth Observatory first gave this theory the official blessing of
the scientific priesthood.

The meteorite was eventually bought by collectors Allan
Lang, Ray Meyer, and Marlin Cilz for $78,000. The car was part
of the deal, and it has since become a world-touring exhibit in
its own right. Michelle proved a lot luckier than Mr. England.
She had originally bought the twelve-year-old clunker from
her grandmother for only $100.

✳

The fall of stones from the sky has not always enjoyed a ready
and widely accepted explanation. Witnesses did not always
have the advantage of the scientific understanding of our universe with which to interpret the experience. Five hundred
years and three days before Michelle Knapp's vehicle unexpectedly appreciated in value by 780 percent, and just sixteen
days after Columbus first set foot in the New World, another
large stony meteorite fell near the Alsatian town of Ensisheim,
France, on November 7, 1492. It is the earliest recorded fall in
the Western world for which the meteorite still survives; it can
be seen today in the town to which it has since brought many
pilgrims and which endeared itself to an emperor during a
particularly turbulent period in the history of fifteenth-century
Europe.

People interpret the world in terms of what they know,
starting with whatever seems most familiar and hence most
likely. In a way, this approach is akin to the geologist's central

doctrine of uniformitarianism, by which the remains of the past are interpreted, wherever possible, in terms of processes we observe going on around us today. After all, it is hardly surprising that a suspicious Leicestershire citizen and a police officer would think immediately of foul play when a rock lands on a car. Policemen encounter crime every day.

Such factors, as well as cultural norms and beliefs, strongly color the way observers report meteorites. Although Michelle Knapp bucked the historical trend in mentioning a smell of sulfur, there did come a point in Western history when reports of meteorite falls ceased to mention it routinely. This is nothing to do with any change in meteorites, of course—and many of them do contain abundant sulfur. What it marks is the decline of brimstone-flavored Christianity.

Despite being half a millennium old, contemporary accounts of the Ensisheim fall are quite precise. It was a little before midday that November 7 when a mighty explosion, heard over an area estimated at 40,000 square miles, announced the fall of a large and roughly triangular stone into a field outside the city walls. A boy in a neighboring field saw the event and running to the scene discovered a hole about three feet deep. Others from the town soon joined him, and together they dragged the meteorite from its crater.

If you visit Ensisheim today, you can see the spot where the meteorite fell at an unremarkable intersection called Les Octrois, little more than a T-junction not far from a bend in the road leading northward to Ensisheim itself. Ensisheim sits on the vast Rhineland plain separating the nearby Vosges to the west from the Alps to the east; the land around it is completely flat. Although trees today obstruct any view of Ensisheim from Les Octrois, it is barely a half mile away.

By contrast with Barwell and Peekskill, few accounts of any associated fireballs survive, even though much of southern Europe must have witnessed the meteor streaking northwestward. A handwritten parchment dated 1513 by Diebold Schilling depicts the fall with a painting in ink and tempera, including a blazing reddish cloud out of which the gray stone emerges. The town of Ensisheim rears up behind, with the Vosges mountains in the far distance and some fictitious hills in the foreground being harrowed by two fictitious laborers.

A more intriguing depiction was made by Albrecht Dürer, who in November 1492 was staying at Basel, only 25 miles to the south. This unsigned painting was found on the reverse of another unsigned (but undisputed) Dürer image of the penitent St. Jerome. And as Ursula Marvin, doyenne of historians of meteoritics, has written: "The subject is neither a star nor a comet. In fact there is nothing it could be, except an exploding meteoritic fireball." Marvin also reports that two art historians believe that this painting, now in the National Gallery in London, is Dürer's depiction from memory of the Ensisheim event. It has been given the title *A Heavenly Body* and is dated c.1495–6. Dürer's later and much more famous engraving *Melancholia* (1514) also carries what is often described as a comet.

Comets are icy bodies that originate far outside the orbits of the planets, but which occasionally become deflected into orbits that bring them to within sight of Earth. As these bodies approach the Sun they become visible because they develop a tail. A comet's tail does not trail behind it in its flight, as would the smoke of a glowing ember thrown through the air; but always points away from the Sun—as Chinese astronomers first realized as early as the sixth century. This is because the tail consists of material blown from its surface by the solar wind, a highly energetic stream of particles and radiation that flows out from our local star in all directions. Anyone who has witnessed one of the recent comets to come near the Earth, such as Halley or Hale-Bopp, will know that they hang in the sky for days or weeks and have an almost serene appearance. The violent image in Durer's 1514 engraving, by contrast, is clearly not that.

Albrecht Dürer's Melancholia.

Back in Ensisheim, an ugly mob was busy smashing bits off the meteorite when the regional governor arrived and called a halt. He ordered the stone (which now weighed about 298 pounds) to be taken into Ensisheim and placed by the door of its church. And this is the point at which the Ensisheim Thunderstone (as it was named) began to play its role in human history.

Although 1492 is today remembered for Columbus's arrival on the island of San Salvador, the people of Ensisheim had other matters on their minds. This area of Alsace has been hotly contested by France and Germany for centuries, almost to the present day, though in the fifteenth century Germany did not exist as we know it today but fell within the borders of the Holy Roman Empire. The public interest generated by the Thunderstone soon came to the attention of Maximilian I of Hapsburg, son of the Holy Roman Emperor Friedrich III, with whom Maximilian was then ruling jointly.

Maximilian arrived in Ensisheim fifteen days after the fall. As kings will, he asked his advisers to interpret the event; and as advisers will, they told him it was a token of God's favor and a sign of good fortune in his continuing conflicts—with Turks in the south and east, and more immediately, with assorted French in the north and west. Suitably pleased, Maximilian had two chunks hacked off the Thunderstone—one for his own cabinet of curiosities, and another for his friend and second cousin, Archduke Sigismund of Austria. Then, on his orders, the miraculous stone was chained up in the church choir-loft to stop it shooting off again, should it have a mind to. Luckily for his advisers, subsequent events did indeed go well for Maximilian, who won the impending battle of Dournon, and signed the Treaty of Senlis in 1493, enlarging his dominions with the addition of Artois, Franche-Comté, and Charolais.

Having served its PR purpose, the Ensisheim Thunderstone remained for the next 301 years chained in its choir-loft.

Surviving even the waves of invasion and counter-invasion of the Thirty Years War (1618–48), it remained safely disregarded until 1793, when the French revolutionary government placed it on display in the Bibliothèque Nationale at nearby Colmar. More pieces were then chipped off, including one weighing about a pound for a certain scientist by the name of Ernst Chladni, who was to revolutionize the scientific understanding of meteorites and is widely regarded today as the father of meteoritics.

By 1803 the Thunderstone was back in its original ecclesiastical resting place, where fifty-one years later it survived the collapse of the bell tower. In 1863 the present church opened on the site of the old; and the meteorite made its final journey, across the square (now a parking lot, inevitably) to the Hôtel de la Régence, then the seat of Hapsburg government. It is now the centerpiece of the Musée de la Régence, which today occupies the building.

It is a rule of life that people in charge take a dim view of those who bring them bad news, so we need not look far for an explanation of Maximilian's advisers' upbeat interpretation of the Thunderstone of Ensisheim. However, meteorites are more usually (perhaps understandably) interpreted in terms of evil than of providence—which makes this pro-Maximilian version rather surprising. How did the emperor's propagandists manage this deft piece of PR manipulation?

The concept of an evil omen was not a simple one in fifteenth-century Europe. Indeed, historians believe that at that time evil portents were often allied to longer-term hopes. Medievalist Philip Soergel of the University of Maryland has written: "While such events were often trumpeted as divine calls to repentance . . . they were almost invariably treated as

hopeful signs that a blissful millennium was soon to begin."
Maximilian had many political supporters who were happy to
promote the advisers' official line, using the Thunderstone as
a news peg on which to hang glowing propaganda about the
soon-to-be emperor.

The Ensisheim fall was the first to take place after the new
technology of printing had become widespread. Not far away,
in the great Alsatian city of Strasbourg, Johannes Gutenberg
carried out his first experiments in movable-type printing, and
by 1440 had already published his results in his book *Kunst
und Aventur* (*Art and Enterprise*). Only ten years later the first
presses were in operation, and by 1492 this new technology
was already four decades old.

Unlike Barwell or Peekskill, the Thunderstone fell before
people had accepted a rational explanation of meteorites, and
did so during a time of high political tension, just as a new
communications technology was becoming familiar. Hardly
surprising, then, that the very first recorded European meteorite fall found itself being "spun." Within weeks, Maximilian's
unofficial PR machine, in the person of influential German
humanist poet Sebastian Brant, had dashed off two political
flyers, "On the Thunderstone Fallen in the Year 1492 before
Ensisheim," complete with dramatic woodcuts.

These depictions strongly recall the pictures that illustrate
Brant's most famous work, a long moralistic tale called *Das
Narrenschiff* (*The Ship of Fools*). Brant was a devout Catholic
and a supporter of Maximilian I, and believed fervently in
Germany's divine right to lead the Christian world. It was
for this reason, he thought, that the Holy Roman Empire had
now come into Germanic hands. As he saw it, the Germanic
peoples' destiny was to rise to this occasion by ceasing their

decadent ways and abiding by the strict morality set out in *Das Narrenschiff*. The arrival of the Thunderstone, he wrote, heralded conflicts in which Maximilian would triumph. With a reformed populace under the control of the Holy Roman Emperor, a blissful new German Catholic hegemony would be ushered in.

Brant's heroic verses describe the fall, citing various ancient authors (including Anaxagoras and Pliny) at length. These illustrious authorities were clearly necessary to give his unlikely sounding tale the credibility it needed in order to bear the kingly portents of his interpretation. Among the literate, sheer disbelief in the very idea that stones could fall from the sky was strong even in 1492 and was to become no less so during the next three centuries—relying, as such reports usually did, upon the testimony of ignorant ploughboys, reapers, and other peasants who spent their days outdoors.

Roman honor, German nation
Stand with you, O King most high.
In truth this stone was sent to you.
God bids you heed, in your own land,
That you should take up arms
O gentle King; and lead your army forth.
Let armor ring and cannon roar!
Let triumph resound, and curb
The vaunting pride of France!

The stirring poem, of which this is a small extract, soon spread widely through Europe; but as with all PR, not all those who read it felt disposed to accept the pro-Maximilian interpretation of the Thunderstone. A different woodcut, whose visual imagery casts the event in more sinister tones than Brant would have liked, was included (together with quotations from the Latin version of the poem) in the manuscript of Sigismondo Tizio's massive and chaotic *History of the Sienese Volume VI*, covering 1476–1505. Tizio was a parish priest living and working in Siena. By 1528, the year he finished that volume of his massive work, he had seen the inaptly named Pope Innocent VIII die (also in 1492) and be succeeded amid bribery and corruption by the infinitely worse Pope Alexander VI—otherwise known as Rodrigo Borgia. Tizio had seen the papacy debased, Italy invaded by the French (1494), and syphilis ravaging Europe—little reason to regard the Thunderstone as beneficent.

✳

Determining the contemporary meaning of an inexplicable object or event is far from simple, especially when those events happened hundreds of years ago to people whose worldview differs from ours. The observed fall of a meteorite is, by human standards, a highly unusual event, demanding interpretation. But what constitutes unusual is not a simple concept either. Unusual to us is not the same thing as unusual to the Earth because of the difference in our timescales.

It is perhaps easier to contemplate the meaning of a work of art, by way of illustration. In the notorious sculpture *La Nona Ora* (*The Ninth Hour*) by Italian artist Maurizio Cattelan we see the late Pope John Paul II apparently pinned to the ground by a meteorite, which lies in the crook of his bent right leg. In full regalia, clinging to his crozier, the pontiff wears an expression that could signify pain, fear, or resignation. Shattered glass shards on the red carpet confirm that the projectile has crashed through a ceiling—perhaps of the Vatican itself.

This haunting image (by an artist who is almost as reticent as nature herself) leaves the work of interpretation to the beholder. We can see what it depicts; but what does it mean? Its title clearly refers to Christ's death on the cross at the hands of men. Here, we see his vicar on Earth apparently the victim of—what?

An unbeliever might say a natural accident. Those who look at nature and see the actions of God may come to a more provocative conclusion. Not surprisingly therefore, many religious people have interpreted the artwork as blasphemous. When it was exhibited in John Paul II's native Poland (Zacheta Gallery, Warsaw, 2000), two members of the Polish parliament attempted unsuccessfully to remove the "meteorite." Although the object is not a real meteorite, it is very heavy. This protest drew more attention to the work, which soon became a public scandal. Anda Rottenberg, the gallery's curator, became the subject of anti-Semitic jibes and was eventually forced to resign.

Atheists, on the other hand, might well interpret the sculpture's meaning in other ways. Meteorites and fireballs streaking across the sky are ready metaphors for inspiration and it is little known outside astronomical circles that the Vatican has an astronomical observatory—which itself owns a prestigious collection of meteorites at the Vatican Observatory Museum. Perhaps the pontiff has suddenly been struck by the crushing idea (for him) that the cosmos has no divine creator after all. Clinging desperately to the lever of his crozier, the pope lies pinned, his expression depicting a fading hope that his God can deliver him from the ugly weight of doubt.

An Earth scientist—believer or not—might reflect that although meteoritic material falls to Earth every day, Cattelan's vision depicts an event that appears on the face of it to be extremely unlikely. His Holiness statistically is no more unlikely than any other member of the human race to be killed by a meteorite—odds which scientists have recently put at 1 in 720,000. However the fact that nowadays at least there is only one pope at any given moment emphasizes this unlikelihood. One seems to be witnessing something so unusual that it should never really happen at all. Two highly unusual objects have coincided most strangely, and it is here that *La Nona Ora*'s power as an image lies.

When Thomas Hardy wrote his lament for the sinking of the *Titanic* called "The Convergence of the Twain," he envisaged two parallel creation stories—of the vainglorious ship of the White Star Line, lying wrecked at the bottom of the Atlantic, and of the formation of the iceberg that was its nemesis. Even as the mighty vessel was being put together, rivet by rivet, in Harland and Wolff's Belfast shipyards, the chunk of glacier that would destroy it was calving off in the distant reaches of frozen Greenland. No one could have guessed how these two objects would one day unite in a single epochal event that would help put an end to the nineteenth-century overconfidence that vaunted human superiority over nature. Any number of inconsequential histories might have been written about these two objects. But their unlikely convergence did happen, "and jarred two hemispheres." It is natural to search for meaning in such great events; yet perhaps nothing is more disturbing than to find no meaning at all.

So far, in the 2,000 years since St. Peter first arrived in Rome, no pontiff has met with the fate depicted by Cattelan. In fact, there is no reliable record of *any* human being ever having been killed by a meteorite—though there have been substantiated near misses. For example, the same year that Michelle

Knapp's car was trashed, a young boy playing soccer in Mbale, Uganda, was hit on the head by a fragment of a meteorite that broke up about 16 miles above the Earth and showered the region with about fifty stones, varying in size from 1 gram to over 60 pounds. Luckily it was only one of the smaller ones—weighing about 3 grams—that fell on the boy after bouncing through the foliage of a banana tree.

More seriously, on November 30, 1954, a meteorite weighing almost 9 pounds crashed through the roof of a timber frame house on Oden's Mill Road, in Sylacauga, Alabama, demolishing a radio and hitting a plump housewife by the name of Ann Hodges as she took a nap on her couch under some fairly substantial quilts. Despite the padding she was badly bruised on hand and thigh; but apart from making the lead news story in the December 13 issue of *Life* magazine (the cover featuring Pope Pius XII) the worst insult that Hodges suffered was the discovery that because the house was rented, she and her husband Hewlett had no claim to the celestial stone. Many years before, U.S. law had established that the owner of any meteorite is that of the land on which it falls. Mr. and Mrs. Hodges' landlord, one Birdie Guy, only gave up the claim after a lengthy legal battle. The Hodges decided to buy the meteorite and paid $500 for it. Sadly by that time interest had waned, and nobody else wanted it any more.

When asked how she felt about being hit by a meteorite, Ann Hodges replied with admirable objectivity, "bruised." Her husband Hewlett, on the other hand, imagined himself on the brink of great riches. Yet, in the end, it brought only disappointment and a broken marriage to the unfortunate couple. Ann herself never truly recovered her equilibrium, became an invalid, and died of kidney failure aged only 52. Both she and her husband are said to have wished that they had never been singled out on the night that stars really did fall on Alabama.

Meteorites have never caused any reliably recorded human deaths; there are not even any reliable accounts of animals dying in this way. Even the tale about a dog in Nakhla, Egypt, supposedly vaporized on June 28, 1911, by a meteorite that was later proved to have come from Mars (splashed off the Red Planet by another impact), is probably false—invented merely to claim a slice of notoriety for a neighboring village.

But as human beings, we are programmed by evolution to search for meaning in events and to interpret them chiefly in terms of ourselves. Our first instinct is to ask not "What caused this?" but to leap straight to the question: "What does this mean for us?" How can this event affect us and our chances of celebrity, of an insurance payout, of successfully defeating the French, of the Russians—or aliens—invading Alabama, or of surviving the race for life? Rare catastrophic events always have this power to pull us instinctively toward the future. Pre-scientific observers hoped to receive simple answers to such complex and largely unanswerable questions. The futures of none of these people, in Barwell, Peekskill, or Ensisheim, could ever really have been divined by inquiring of the meteorite—though there is much that can. Instead of asking why, and hoping for prophecies, science demands first of all to know how. Science studies the present, to receive answers about the past; and only by learning about the past does it discover, in the end, the only true insight into what may be to come.

Since it first began to apply itself to things that dropped from heaven, science's approach to meteorites has brought us incredible knowledge and raised many more questions about the origin of the Solar System, our planet, and life upon it. We normally think of finding out about space by going there. In meteorites, however, deep space and deep time come to us, reminding us that, although the Earth may exist in the vacuum of space, it is not isolated from its cosmic environment.

Geology as a science began to make real progress at about the same time as meteoritics did, at the end of the eighteenth century—and did so by ceasing to think in terms of fantastical and picturesque deluges, disasters, catastrophes, and revolutions, turning instead to the slow, incremental, and above all observable processes occurring around us.

And here we come back to the difference in scale between ourselves and our planet, and the notion of what constitutes a rare event for a human as opposed to the Earth. In this context we must also remember that even if an event is rare, it may well be severe enough to cast a long shadow historically, as the fact of its occurrence changes everything that comes after.

Earth scientists now realize that the events we humans witness during our brief lives or record in our fragile civilizations cannot possibly be taken as representative of the whole of deep time since the formation of the Earth. The planets, meteorites tell us, first formed 4.567 billion years ago. Earth today experiences truly great cataclysms so rarely that we—even as a species—have never witnessed one and, maybe if we are lucky, never shall. Some of these events, like the arrival of a gigantic meteorite 6 miles in diameter, can re-shape the whole globe forever but may only happen every 100 million years or so. This interval is easily ten times longer than our species' entire existence. By our standards, such impacts are incredibly rare events. Yet, to the Earth, an event that happens even every 100 million years is comparable, in human terms, to an annual dental check for a 45-year-old: not really that unusual at all.

Given enough time, just as one of an infinite number of typing monkeys will eventually write *Hamlet,* one day even a pope will eventually be struck and killed by incoming cosmic debris. Possibilities are all but endless in the near endlessness of geological time, and meteorites have helped science come to terms with these ideas more than any other natural phenomenon.

People will always interrogate the skies and the things that fall from them just as gallery-goers will query the meaning of an enigmatic artwork. They will develop answers that accord with their view of life, their hopes and fears, and their historical context, and such answers will, or may, be meaningful for them and their contemporaries.

Science's approach, however, to its subtly but significantly different question of how, develops answers that, measured against the testimony of nature, apply equally to all. The study of meteorites has proved the most fruitful of all in finally answering the fundamental question of where and how and when the stuff that we are made of was first created.

2

STARRY MESSENGERS

Ere time began, from flaming Chaos hurled
Rose the bright spheres, which form the circling
world

ERASMUS DARWIN

It is just possible that an alien astronomer, living now in the quasar galaxy we call 3C48 in the constellation of Triangulum, may be staring at our own galaxy with a very powerful telescope and witnessing the moment that our Solar System was born.

If you live in the northern hemisphere, at night you can see the constellation Triangulum, bordering the constellations Andromeda, Aries, Pisces, and Perseus. It consists of just three stars with unmemorable names, belonging to our own galaxy, the Milky Way. But you are not likely to see 3C48 beyond them, hidden in what astronomers call the deep sky. Like a distant object glimpsed through leaded lights, 3C48 is a very faint, radio-emitting object, first identified by radio astronomers and published in the *Third Cambridge Radio Survey* in 1959. Such strong quasi-stellar radio sources were later named quasars; and in 1960, 3C48 became the first of these to be seen through a telescope by human eyes.

It was noticed that 3C48's light was dominated by long wavelengths, toward the red end of the spectrum. Light with such a high red-shift tells of a very distant source, because that light has spent its long life traveling through space that is continually expanding. Space doesn't just expand at the edges—it expands everywhere; so light traveling through space can be thought of as a wavy line drawn on a relaxed elastic belt. As you stretch the belt, the line's wavelength increases.

3C48 was not like the Andromeda galaxy, which sits beyond the Andromeda constellation in the same way that 3C48 is glimpsed through Triangulum. Andromeda is only 2.5 million light years from us, the nearest spiral galaxy to our own. 3C48 proved to be 4.5 billion light years away.

What this means is that we are looking at 3C48 and seeing it as it was long before our Sun and our world even existed. By the same token, intense light from the supernova event that may have triggered the formation of our Solar System within a cloud of gas and dust, creating the Sun and all its planets, will only now be arriving at 3C48. As you stare into space, you stare into deep time. From out of that abyss, telling us how first the universe and then our Solar System originated, come electromagnetic energy, and meteorites. Our story must begin, therefore, with the story of matter itself—of the chemical elements—for they too were not created in a flash, but have evolved through time.

Even the least scientific of us should remember the periodic table of the elements—that rather daunting icon displayed in every chemistry classroom or laboratory, its single and dual letter-symbols sitting like perfect specimens in neat boxes. We tend to assume that this is yet another useful but dull classification of things that are today just as they have been since the dawn of time, but no. Just like Darwin's species, they are telling a story—a story about how the universe has changed since its creation in the Big Bang over 13 billion years ago.

Counting along the rows, after hydrogen and helium, from top left to bottom right (perhaps recalling those mnemonics we may have learned: *Little Beggar Boys Catch Newts or Fishes; Naughty Maggie Always Sips Port, Sherry, Claret*), we are climbing a ladder from the lightest elements to the heaviest. The process stretching back behind this set of boxes is perhaps the grandest natural process of all; for, as stars burn away they produce—and then recycle as new fuel—progressively heavier chemical elements. Gradually, as more and more generations of stars have formed, lived, and died since the birth of the universe, the process of stellar nucleosynthesis (which scientists first understood in the mid-1950s) has seeded the universe with the matter that eventually went into making worlds— and us.

It was only a decade or so after this epochal discovery was first made that the head of chemistry at my school one day found himself substituting for our regular teacher and facing a first-year class—something to which he had long since grown unaccustomed. Schools then were not as hung up about the curriculum as they are today, and so instead of trying to replace our lost lesson, Graham Gregory stared for a few moments at the periodic table, then launched into an impromptu lecture on the life and death of stars.

When he got carried away, Gregory was in the habit of throwing one arm over the top of his bald head, so that his left palm covered his right ear; and as his lecture ended, he found himself in this pose, eyes shut tight, like a preacher. "Gentlemen," he concluded, "I would ask that you allow yourselves to be struck by the notion that the simple presence of elements like iron, silicon, and carbon on our planet and in ourselves means that since the birth of the universe, stars, perhaps with planets like our own, lived and died long before our Solar System condensed from their exploded remains, and of which our heavier atoms are now the only surviving trace."

✳

At 3:00 p.m. on November 2, 1936, the BBC television service began broadcasting for the first time. The opening show featured, for the benefit of the 400 or so wealthy viewers who witnessed it, a song called "Television" with lyrics by writer, composer, and actor James Dyrenforth. Sung by Adele Dixon, the song concluded with the lines:

> *There's joy in store*
> *The world is at your door—*
> *It's here for everyone to view*
> *Conjured up in sound and sight*
> *By the magic rays of light*
> *That bring Television to you.*

This may sound patronizing to our ears, but the physics is sound. Television is brought to us by radio waves, which are a form of electromagnetic radiation. Electromagnetic radiation includes visible light as part of its spectrum. Just as the sound spectrum includes frequencies that are too high or low for us to perceive, so does electromagnetism. But whether the radiation in question is ultraviolet, visible, infra-red, radio, or microwaves coursing through your frozen lasagna, they are all rays of light. The only error in the song is that there's nothing magic about them.

The universe is therefore flooded with light, not all of it visible to us. Among the most invisible light is the oldest (and hence most red-shifted) in the universe, created during the Big Bang. Astronomers call it the cosmic microwave background—the after-glow of the most energetic event in the universe's history, its birth. From its degree of red shift, astronomers calculate that the universe is around 13.7 billion years old.

The Big Bang, which as everyone knows set the universe going, cosmologists say actually took three minutes, as all the subatomic particles (roughly half matter and half antimatter) were created at almost infinite density and began the expansion and cooling that continue to this day. Matter and antimatter then canceled each other out—except for a slight imbalance in their ratio, which left the universe with all the leftover matter that it now contains.

At the end of time's first 180 seconds, all the matter in the universe consisted of the first three elements of the periodic table—hydrogen (c.77 percent), helium (c.23 percent) and one millionth of a percent of lithium. The mixture thinned out as the universe expanded, first becoming transparent after about 300,000 years. But as Lewis Carroll's Walrus and the Carpenter might have observed, no stars were shining in the sky—there were no stars to shine. After 300 million more years, the celestial fog began to condense—forming a network of filaments. Within these, the first stars, clustered in the first galaxies, first let there be light.

Stars are nuclear furnaces within which atoms crush together and fuse into heavier elements. Nobody has ever observed one of these very early stars, even in the most distant reaches of the universe. Yet, as they formed and died, the richness of the universal chemical brew increased. Ever since then, the amount of heavy elements throughout the universe has everywhere increased, thanks to the process of nucleosynthesis. Yet despite that a star forming right now will form from much more chemically mature raw materials, it will still principally burn hydrogen. Our Sun, for example, converts 616 million tons of hydrogen into 611 million tons of helium every second, losing five million tons of mass in the process, as energy, according to Einstein's famous ratio whereby energy (E) is equal to the mass (m) times the speed of light (c) squared, or multiplied by itself. Even though it is 4.5 billion years old, the Sun has so far only managed to convert about 4 percent of its original hydrogen into the heavier element.

Not all stars are the same, and as with people, how massive they are will strongly affect the kind of life they will lead and the death they will suffer. For the purposes of knowing where the chemical elements came from, and how they came to build our Solar System, we need focus only on one sort of star-death—the supernova. There are two basic types of supernova, imaginatively called type I and type II, each delivering slightly different proportions of heavy elements. Both have had their part to play in contributing their dusty ashes to our Solar System. In fact cosmochemists believe that between thirty and forty different star deaths contributed matter to the dust cloud from which our Sun finally condensed. The Sun, however, is too small a star to end in a supernova. For this to happen, and to produce elements heavier than carbon, oxygen, and nitrogen, a star must be at least four times bigger, and ideally between eight and ten times bigger, if it is to create elements further up the periodic table than iron.

Nucleosynthesis was first revealed to the world in one of the two greatest scientific events of the 1950s (the other being Watson and Crick's discovery of DNA) and perhaps one of the greatest scientific papers ever written. The paper was so groundbreaking that it is known familiarly as "B^{2}FH," a jocular formula created from the initials of its four authors—arranged alphabetically: Geoffrey Burbridge, Margaret Burbridge, William Fowler, and Fred Hoyle. While all authors properly shared the credit, there is little doubt that the inspiration and most of the initial work were by Fred Hoyle; though it was Willy Fowler who was singled out later by the Nobel committee for their physics prize.

In the late 1940s and early '50s, Fred Hoyle was returning to academic life after wartime work on radar. He had already built up an idea about what actually was happening when a super-massive star explodes in a supernova—liberating so much energy that for a short time it outshines all the stars in the galaxy. Hoyle knew that atomic nuclei fused when gravitational pressure overcame the electrical repulsion between them, and that they did so because it enabled their nuclei to achieve a more favorable energy balance than by remaining apart. In the Sun, hydrogen fuses to form helium. When the hydrogen supply fails, the star begins to burn helium; and so on, up the periodic table. Thus the process of nucleosynthesis creates bigger and bigger atoms including carbon and oxygen—as far as iron, element 26.

At this point the process in smaller stars hits a wall. Beyond iron, much more energy is required to force atoms together—which is why it can only be achieved in a star of truly colossal mass. This, Hoyle thought (and with the help of B^2 and F eventually proved), was the trigger for a supernova, when stars above the critical mass hit the iron barrier, collapse under their own gravity, and blow themselves to bits—seeding the galaxy.

A supernova is brilliant enough to be seen from another galaxy; so we can imagine our alien astronomer on 3C48 noticing one, lighting up a region of the distant Milky Way, near a cloud of gas and stardust. This cloud, a stellar nursery, was itself derived from the death of massive stars during the nine billion years or so that had by then elapsed since the great darkness ended and stars first lit the heavens.

It was the shock from that supernova, most scientists believe, that triggered the initial gravitational collapse of the molecular cloud that, after about 100,000 years, gave birth to an energetic early Sun surrounded by a turbulent, dusty disk spinning around its central star. As the star grew and caught light, it swallowed all but about 1 percent of the total mass of the collapsed cloud; but that 1 percent was enough to form, eventually, all the planets and other bodies in the Solar System.

The inner parts of the rotating disk, being closer to the Sun, were hotter than the outer regions, and here the rocky planets (and the asteroids) would one day form. Beyond them, temperatures were low enough for water and other lighter molecules to condense, and beyond this so-called snowline the planets would all be gas giants, consisting of light elements as well as molecules like water, carbon dioxide, methane and other volatiles. Beyond them would form a zone of shadowy, icy bodies called the Kuiper Belt, of which Pluto is king. Even further out (though as yet unobserved), at the far edge of the Sun's gravitational influence, float billions of icy objects—potential comet nuclei—in a zone known as the Oort Cloud.

Despite these differences created by the temperature gradient, the bulk chemical composition of the Sun and its accretion disk were uniform—all displaying the so-called solar composition inherited from all those supernova events that added to the original brew of gas and dust from which everything condensed. It is a chemical signature that we all share, and that speaks to us of our galactic place like the smell of home.

We can see the chemical make-up of the Sun in sunlight. The solar spectrum carries a signature showing which elements are present and how much there is of each. This is useful because planets and asteroids may undergo processes that segregate out certain elements into different types of rock and metal. The degree to which a rock has been differentiated and hence moved away from its original solar composition is a measure of how much geological processing has gone on subsequently—and may also indicate what those processes were.

In its early life, the Sun was rather different; and most astronomers believe that it was probably what is known as a T-Tauri star. This type of star takes its name from the type specimen in the Hyades Cluster in the constellation of Taurus. It is very young—possibly as little as one million years—and was discovered in October 1852 by British astronomer John Russell Hind, who was also a prolific early discoverer of asteroids. (He found eleven, one of which he named Victoria, a controversial move because naming things for living persons, even royal ones, was not done in those days.)

T-Tauri stars (and many more are now known—a recent study looked at 500 in the Orion Nebula alone) are on their way to achieving the required mass for the hydrogen fusion process to begin—which is why they are all younger than about 10 million years. Such stars often have large accretion disks around them, and they glow, not by the steady light of nuclear fusion, but by the fitful conversion of gravitational energy—and for this reason are also known as variable stars.

The solar wind began to blow, driven along magnetic lines of force. This prevented further accretion of material to the Sun, opening up an annulus of clear space between the star and the inner edge of its encircling disk. Meanwhile, by tugging on the proto-star's magnetic field—which, being unable to penetrate the disk, squeezed through the clear space like a napkin through a ring—the disk exerted drag—slowing the star down and transferring its rotational energy to the disk. This is probably why the Sun spins so slowly, completing one rotation in a leisurely 28 Earth days.

Eventually, continuing gravitational collapse triggered the fusion of hydrogen atoms into helium. The Sun's first steady light would have revealed, circling the new star, a swirling mixture of dust and gas—which was slowly curdling. Clumps orbiting close to the Sun and heated to higher temperatures would one day form the rocky planets and asteroids. The outer clumps, beyond the snowline, where volatile substances like methane and water were stable, formed the gas giant planets—Uranus, Saturn, and, largest and innermost, mighty Jupiter—whose deep gravity well both protects the inner Solar System from potential impactors by sucking them up like a celestial vacuum cleaner but yet also sends rocky objects—the meteorites—from the Asteroid Belt, and in toward the Sun and our home planet, connecting us with the pre-geological history of the Earth.

But there are many problems to overcome in understanding these time capsules from before geology began. A seismologist who uses seismic waves from earthquakes and explosions to x-ray the Earth and understand its structure, once described his subject as like working out the structure of a piano by listening to one being pushed downstairs. In the same way, deciphering the early history of the Solar System from meteorites—our only material link with those times—can be a little like trying to work out the news of the day after your newspaper has been turned into papier mâché.

Apart from ready-made stardust (smoke and ashes of old supernova explosions), the other solid matter in the disk had to condense as it cooled, in a regular and well understood sequence. If seawater is allowed to evaporate away in a dish, the most insoluble of the salts it contains precipitate first and the most soluble salts last. In a similar way, as the protoplanetary disk cooled, the first chemical elements to precipitate were those that become "insoluble" at the highest temperatures.

When these elements condensed, they formed the oldest un-inherited matter—they were the Solar System's very first homemade stuff. Many of these fluffy specks of whitish calcium and aluminum compounds occur in meteorites and they have yielded the oldest radiometric dates ever obtained—4.567 billion years—the memorable figure now taken as the benchmark age of the Solar System. No longer was the darkness without form and void.

These calcium-aluminum inclusions (CAIs) may have formed right at the inner edge of the disk, where temperatures were highest, and were carried upward and outward by a solar wind—showering the nebula like a garden sprinkler. This idea was boosted in 2006 when material brought back by the Stardust Mission (which collected dust from the trail of comet Wild-II) was found to contain CAIs. Comets are believed to originate in the most distant reaches of the outer Solar System, right at the edge of the Sun's gravitational sway, in the distant Oort Cloud. Because of the high temperatures at which they condensed, it is extremely unlikely that any CAIs found in comet dust could have got there without having been carried there.

CAIs formed during the first 300,000 years of the disk's existence, by the end of which time the nebula would have thickened a little to contain three components—CAIs, stardust, and more gas. The mystery deepens when we ask what happened next. For it is clear that a very great deal has happened between the formation of CAIs in a dusty nebula, and the Solar System of today. The meteorites are our witnesses to that span of time, when the rocky planets Mercury, Venus, Earth, and Mars grew.

As the solar nebula cooled to nearer 1,500 degrees Celsius, silicates that build minerals familiar to geologists—like olivine and pyroxene—formed. Grains of metallic iron and nickel also appeared in the mix. Iron reacted with sulfur to form iron sulfide. Lastly, at nearer to 500 degrees, the more volatile minerals condensed—the sulfates, carbonates, clay minerals (silicates with water bound in their crystal lattices), and even complex molecules built out of carbon.

The nebula's temperature gradient ensured that the innermost part of the disk became richest in the refractory minerals and metals. The innermost planet, Mercury, formed from material orbiting closest to the Sun, has the largest iron core of all four rocky planets. Mars, by contrast, has the smallest core and the lowest overall density.

✳

So what did this primordial solid matter, from which all planets, asteroids, and meteorites formed, actually look like? We have all seen footage of astronauts in zero gravity drinking by allowing water to spill into the cabin and then catching the wobbling blobs in their mouths as they float by. Liquids in space form spheres—there is nothing else they can do—and that is part of the problem in understanding the main components of the majority of meteorites. So when material becomes molten in space, it forms a sphere, cools, and solidifies into a spherule which meteoriticists call a chondrule.

After CAIs, some chondrules formed by the melting of pre-Solar dust and so constitute the most primitive solid matter in the Solar System. If you blow powdered sugar into the hot gas escaping from a blowtorch, and if you get the temperature just right, the sugar will not burn but caramelize and rain out on to your work surface as brown beads. That is essentially how chondrules formed, as a fiery rain—to use the words of the Sheffield geologist and pioneering microscopist Henry Clifton Sorby, who was among the first to see and describe chondrules under the petrological microscope. About 90 percent of all meteorites that fall to Earth are made of chondrules—they are called ordinary chondrites for this reason, though it is a strange name to give to such extraordinary objects, about which there is more disagreement and controversy than almost anything else in meteoritics. The problem is that nearly every kind of event that melts material in space will create chondrules, and in their immensely long history, those that formed the earliest will probably have been affected by other processes since.

Did the first chondrules form as dust spiralled in toward the early Sun, and did they then become caught in a hot solar wind like powdered sugar blown across a flame, which singed, melted, and blew them outward across the disk? Were there perhaps mysterious shock waves emanating from the Sun, whose energy turned to heat and melted the dust? Or did they form in hypervelocity collisions, when accreting bodies smashed into one another and were vaporized? Were there powerful electric currents arcing through the dust like lightning through clouds, whose strikes flashed the dust to liquid?

Because all melting events affecting the materials of the nebula are going to form the same product—glassy beads of fiery rain in the weightless vacuum of space—it is exceedingly difficult to find ways of eliminating any of these ideas. Perhaps, indeed, all had some role to play. Most recent analyses seem to imply that chondrules formed in flash heating events of some kind—which will have re-melted them, perhaps more than once. But what caused these events is very uncertain and hard (and probably wrong) to attribute to any single cause.

But we do know that chondrules began to form a little over one million years after the calcium-rich inclusions condensed, and continued to form for several million years thereafter. They are the original cosmic sediment; but at some stage this sediment must have started to come together to form bigger objects. This process of accretion, which ultimately created the planets and asteroids from the raw material of the protoplanetary disc, brought with it changes that radically affected the texture and chemical composition of the material comprising the larger bodies.

The early Solar System contained much more natural radioactivity than it does today. This is because radioactive elements eventually decay into various non-radioactive products, at a rate expressed in half-life, the time it takes for half of the total amount of a radioactive substance to be used up. Half-lives of different radioactive elements vary enormously in length—from fractions of seconds to billions of years. However, the nebula contained many short-lived radioactive isotopes that now, after the passage of so much time, have become completely extinct. With all this extra radioactive decay going on, more heat was generated. So as soon as the first planet-like bodies began to form, they got a lot hotter inside than they would if they were to accumulate from the dust of the modern Solar System. One important such isotope was aluminium 26. This unstable form of the familiar metal would have been contributed to our nebula by a supernova—perhaps the one that astronomers think began the whole process of Solar System formation, and which we can imagine being witnessed today, 4.5 billion light years away in that quasar galaxy 3C48, in the constellation of Triangulum.

As bodies with diameters of a few hundred to a few thousand yards in diameter began to accrete, they would quickly have become very hot and melting would have ensued, the temperature being due to gravitational collapse, condensation, and these active but short-lived radioactive isotopes. As a result, the interiors of these early planet-like bodies, called planetesimals, would have begun to melt and separate out, or differentiate. Native metal (nickel and iron) is immiscible with and denser than rocky silicate magma, and so would have run together and sunk to the core, taking with it all those elements that dissolve more easily in liquid metal. Rare elements like iridium and osmium would thus have been progressively leached from the surface. (Geochemists call these elements siderophile because they "love iron.") The disappearance of iridium into the cores of planetesimals and later of planets will have profound significance later in our story.

Surrounding the planetesimals' cores, the silicate mantles would have been molten too, and the resulting magma may then have erupted, forming crusts of solidified lava. Small impacts would have pulverized these crustal rocks, pocking them with craters and generating regolith (the accumulated ejecta of many impacts), in a process of surface reworking that astronomers call gardening. This regolith might at that point become fused into a rock known as a breccia. Bigger impacts might then totally disrupt the body, liberating even fragments of its cold, dead, metallic heart into space—one day to fall to Earth, perhaps, as iron meteorites.

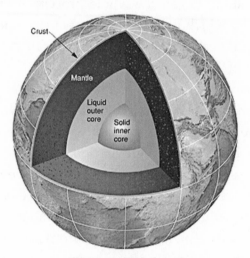

Core, mantle, and crust

This threefold structure—core, mantle, and crust—may remind you of the structure of our own planet, which to this day has a core of nickel-iron, a mantle of solid but hot and plastic silicate minerals, and a thin crust. Such a structure is common to all rocky planets like ours—the Earth being still geologically active after all this time purely because it is bigger than its neighbors. The planetesimals could not maintain active geology because their internal heat source gave out as the amount of aluminum 26 declined.

If you remember some basic school math, you might recall that the volume of a sphere increases much faster than its surface area; so it follows that surfaces of big spheres like the

Earth are smaller in proportion to their volume than those of small spheres like the Moon. The amount of radiogenic heat produced inside a sphere is proportional to volume, but the heat lost to space is proportional to surface area. Thus a large sphere will be able to hold in its heat much longer than a small one. This is why the Earth remains geologically active, and why smaller objects like Mars or the Moon solidified hundreds or even thousands of millions of years ago.

It is likely that the four rocky planets (Mercury, Venus, Earth, Mars) accreted to their current sizes after about 100 million years. In the cooler regions of the nebula beyond the snowline, bodies formed that consisted of pristine solar nebula material mixed with water and other volatile substances such as complex carbon compounds, similar to those originally known from the processes of life on Earth, and which chemists therefore call organic. Although these relatively soft, delicate meteorites (the carbonaceous chondrites) are rare finds on Earth, they appear to be plentiful in space.

The snowline is the great divide between the inner and outer Solar System. Jupiter has a small rocky core, but it was able to draw in huge amounts of volatiles such as hydrogen and helium that now compose most of its mass. The moons of the great gas giants are themselves mostly mixtures of ice and rock, liberally mixed with methane and ammonia. Jupiter, 317 times the Earth's mass and with huge gravitational power, exerted a profound influence on how the rocky planets grew; or failed to grow—an influence that continues today.

✳

So, apart from the temperature in the protoplanetary disk, which decreasing outward gave us inner rocky planets and outer gassy and icy planets, what else, if anything, determined the planets' eventual position and spacing?

German astronomer Johann Elert Bode was only nineteen when in 1768 he prepared the second edition of his guide to the starry sky, *Anleitung zur Kenntniss des gestirnten Himmels*—a kind of catalog of known celestial objects. In a footnote he added the observation that the planets then known (Saturn was then the most distant known) seemed to occupy orbits that were spaced at regular intervals. He came to this conclusion by dividing the distance between the Sun and Saturn into 100 parts. Mercury sat at 4 units' distance. Venus sat at 7, which is 4 + 3. Earth sits at 10, which is 4 + 6. Mars at 16 comes in at 4 + 12. Jupiter sat at 52, which is 4 + 48, while Saturn sat at 100 divisions, which is (of course) 4 + 96.

The second surprise about this mathematically regular progression, the first being that one existed at all, was that there was apparently a gap in it—between Mars and Jupiter. The mystery of the missing planet seemed to spoil an elegant model. Then, in 1781, the English astronomer William Herschel discovered Uranus. The hairs must have risen on a few peoples' necks when they realized that the orbit of the new planet was precisely where Bode's series would have predicted.

Although Bode was the first to get his name on this "law" it is now known by the joint names of Bode and Johann Titius. The almost spine-chilling regularity of the spacing of planets predicted Uranus, but also demanded the presence of a new planet between Mars and Jupiter. Then, on New Year's Day 1801, Giuseppe Piazzi of Palermo discovered Ceres—at 597 miles in diameter, the biggest of what we now call the asteroids. It was heralded as another successful prediction. Piazzi's honor, however, was short-lived as more "missing" planets started turning up, complicating matters considerably.

In fact, the more accurate measurement of planetary distances began to reveal that the Titius–Bode Law's predictive power was in fact a lot less impressive than had first appeared. Throughout the series, actual and predicted distances were found to coincide only rather approximately. Moreover, although Uranus had been in roughly the right place, Neptune, discovered in 1846, was way off.

Astronomers now think that because matter in the protoplanetary disk was distributed according to fairly simple physical laws, it followed that as the growing proto-planets swept up the material within their gravitational grasp, they were, on average, rather more likely than not to create a roughly regular Solar System, with its constituent bodies following (though not closely) a simple arithmetical distribution. For that reason, astronomers have downgraded Titius–Bode from a law to a mere rule. And as with all rules, it is the exceptions that are interesting—in this case the planet that was apparently "missing" between Mars and Jupiter. While there was indeed no planet in the predicted position, the apparently vacant space was not entirely free of solid matter. The Asteroid Belt occupied it.

We have a rather distorted view of what the Asteroid Belt is like, and Hollywood is to blame. Scenes like the one in *The Empire Strikes Back* where Han Solo evades the imperial fleet by flying the Millennium Falcon into an asteroid field, have given us the impression of chaotically moving, frequently colliding, and rather densely spaced rocky objects.

In fact, the Belt is nothing like that. It extends unevenly between the orbits of Mars and Jupiter, a zone with an inner edge at about 158 million miles from the Sun and an outer at about 598 million. Yet this vast region of space contains today less material than would make one Moon. Countless billions of asteroids exist in the Belt, mostly very small; though more than a billion of them are probably over half a mile across. Yet most probes launched from Earth to the outer Solar System have sailed straight through the Asteroid Belt, and those that have deliberately made asteroid fly-bys have had to be very carefully aimed.

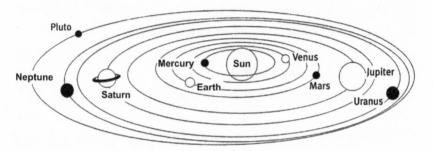

The major planets of the inner and outer Solar System. The Asteroid Belt is located between Mars and Jupiter. Mars is the last rocky planet and Jupiter the first of the gas planets. Pluto is the icy object now classified as the dwarf planet, probably being the largest of the distant Kuiper Belt Objects.

Each traveling along its own independent orbit, these ancient and often irregular, potato-shaped objects spin and tumble and (very rarely) collide; though most of the time they remain thousands of miles from one another. In its early history, the Belt was much more crowded, and astronomers believe that since the time that the other planets formed, the Belt has lost as much as 99.5 percent of its solid mass—the equivalent of about three to five Earth masses.

Titius and Bode were not really wrong about the missing planet between Mars and Jupiter. There is no planet there simply because Jupiter's gravitational influence was always disruptive, never allowing the material in that part of the nebula to collide creatively. When Piazzi discovered the first asteroid, he thought he had discovered that lost planet. However, it was not long before more lost planets came hurrying up and in fairly rapid succession. Asteroid Pallas was found in 1802; Juno in 1804; Vesta in 1807. Astraea had to wait until 1845, but after that they came thick and fast and by 1890, 300 were known. Now, almost 10,000 have been identified and plotted in their orbits. The material of Titius–Bode's missing planet had been there all the time. But that planet had been stillborn, 4.6 billion years ago.

✳

Testament to the continuing influence of Jupiter's gravitational pull is a strange phenomenon that (if it were possible to look upon it from far above or below the plane of the ecliptic) would give the Asteroid Belt a stripy appearance, similar to that of Saturn's rings. Of course, we always see the Asteroid Belt edge-on, so the pattern of concentric rings only emerged on paper as more and more asteroid distances were measured with accuracy.

Careful plotting of asteroid orbits reveals five bands, separated by four narrow zones of clear space. In the graph, these gaps can be clearly seen and occur wherever the orbit of any asteroid occupying that space would experience an orbital resonance with Jupiter—i.e., would complete a certain whole number of orbits of the Sun in the same time as Jupiter would. Thus the gaps occur where the ratio of asteroid years to Jupiter

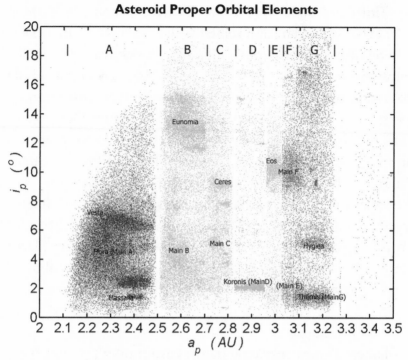

A transverse section of the Asteroid Belt: Mars to the left, Jupiter to the right. The scale along the bottom is marked in Astronomical Units (Earth–Sun distances). Clear lanes represent Kirkwood Gaps, swept clear of asteroids by gravitational resonances with Jupiter. Many asteroids that stray into the gaps may be sent into Earth-crossing orbits. The family labeled Ceres is now known as the Gefion Family, whose creation in a catastrophic collision is thought to have been the source of the Ordovician asteroid bombardment.

years are 3:1, 5:2, 7:3, and 2:1. (There are also less obvious gaps at ratios 8:3, 5:3, 3:2, and 4:3.) So how do these gaps originate?

Think of a notional asteroid propelled, perhaps by some unlucky cannon following an impact with another asteroid, into an orbit within the 2:1 gap. This body would now go around the Sun in precisely one third of the time that it takes Jupiter to make the same circuit. This means it would line up with Jupiter every three "asteroid" years—at which conjunction it would find itself tugged by Jupiter's gravity. Over millions of years, these repeated tugs would tend to pull the object out of the resonant orbit. Of course, all other asteroids in all other orbits also feel the pull of Jupiter's gravity whenever they too happen to line up between the giant planet and the Sun. But those occupying these resonant orbits feel the pull of Jove more often—effectively ruling those particular orbital distances off limits.

Since their discovery in 1857 by astronomer Daniel Kirkwood, these empty lanes have been known as Kirkwood Gaps, and it turns out that they have a profound influence on our planet. As Kirkwood himself wrote, the sorting out of asteroids into distinct bands would have one very important consequence for Earth and the other rocky planets: "If the [original] distribution . . . was nearly continuous . . . it would probably break up into a number of concentric annuli. On account, however, of the great perturbations to which they were subject, [objects in] these narrow rings would frequently come into collision." In fact, the trajectory of an asteroid entering one of the Kirkwood Gaps can be perturbed to such a degree that it may leave the Asteroid Belt completely. Kirkwood Gaps act, in fact, like sliproads leading off the asteroid motorway and into our planet's backyard.

Confirmation that at least some meteorites come from the Asteroid Belt had to wait until the era of mass photography and fast film, because astronomers can only tell where a meteorite comes from if they know its orbit before it fell to Earth. To know that, they need photographs of fireballs entering the atmosphere, just like for the Peekskill meteorite, whose perturbed orbit ended in the trunk of Michelle Knapp's Chevy

Malibu. Gradually, from 1959 on, as more and more meteorite falls have been recorded on film and video, it emerged that these incoming objects had orbits whose furthest point from the Sun lay in or near the Asteroid Belt.

✳

Even in the zones between these clear lanes, which if viewed from above would look like the gaps between songs on a vinyl record, asteroids are not uniformly distributed. Rather, they appear clumped—and each clump is the trace of a great collision, long ago.

Despite what Hollywood might have us believe, when two asteroids collide, they probably do so quite slowly. When this happens it is possible that the fragments produced might possess just enough mutual gravitational pull to prevent the pieces created from flying completely apart. The fragments then pull together again, creating reassembled asteroids that are sometimes called rubble piles. However, collisions will often create a group of bodies that just hang together—continuing in their orbit, like a flock of geese.

So, as well as being sorted by Jupiter into zones separated by the Kirkwood Gaps, asteroids in the past have, by colliding at just the right relative speed, given rise to clusters of closely associated fragments of one or more parent bodies which ever since have been traveling around the Sun together, fading gradually through time as outer members drift apart or are picked off by gravitational interference.

This fact was first realized in 1918 by Japanese astronomer Kiyotsugu Hirayama, who identified the first five groupings that are now called Hirayama families. Circulating through the main Belt near Kirkwood Gaps, they provide samples to the conveyor belt that leads to Earth.

Gradually, astronomers came to realize that asteroids did not stay in the Belt, and that the Earth-crossing asteroids (first observed in 1873, when astronomer James Watson observed asteroid 132 Arethra and noticed that its orbit passed inside that of Mars) also went around in groups, though these were groups that were not necessarily genetically related to one another.

The asteroid 433 Eros was first seen in 1898, with an orbit that took it within 13 million miles of Earth's orbit. In 1938 asteroid 1221 Amor was discovered, with an orbit that took it from just beyond the Main Belt to a closest approach to the Sun only 8.5 million miles from the orbit of our own planet. Since then, astronomers have recognized three categories of asteroid orbiting the Sun in near-Earth space—and named them the Amor, Apollo, and Aten asteroids. Together they constitute the Near Earth Objects (NEOs).

Since astronomers began looking for these in earnest—and started attracting public attention to their work by issuing doom-laden press releases about it—these objects have occasionally made the news. The public watched with relief, though later with increasing annoyance, as each potential threat, first announced with the warning that we might all be sent the way of the dinosaurs, was quickly downgraded to completely harmless—usually within a few days, or as soon as more accurate orbital information had been gathered.

Meteorites whose entry to the Earth's atmosphere has been caught on film in recent years—such as Lost City (1970), Innisfree (1977), Peekskill (1992)—have allowed astronomers to calculate the orbits of their parent bodies. All, it turned out, originated somewhere near the Asteroid Belt—Peekskill on its inner edge. But the Earth-crossing asteroid families like the Amor, Apollo, and Aten clusters are less like families than orphanages. NEOs do not last long in geological/astronomical terms, and just as children eventually leave an orphanage when they are adopted, within 10 to 100 million years all the current population of NEOs will be sucked down into the gravity well of one of the planets whose orbits they cross, and so find an adopted planetary home at last. Thus, the meteorite that wrecked Michelle Knapp's Chevy had in fact been abandoned by its parent—and probably propelled into its Earth-crossing orbit at the same instant—32 million years earlier.

As there are thousands of NEOs, their population must be continually replenished; and for most NEOs, their original home was usually any resonant orbit close to a Kirkwood Gap within the Belt. So, most of the meteorites landing on Earth come originally from parent bodies situated near the borders of, or just within, one of the Kirkwood Gaps. Our quest to locate the genetic parents of these cosmic orphans has led us to their last known address. The next question is—might there be an equivalent of a DNA test that could help us to reunite particular meteorites on Earth with particular parent bodies in the Asteroid Belt?

The answer is probably yes; but to understand how, we must return to nineteenth-century Sheffield and to the achievements of its greatest scientific son, the geologist Henry Sorby, who first examined meteorites under the microscope and described the chondrule.

In 1666, Sir Isaac Newton first realized that the whiteness of sunlight was in fact composed of a spectrum of different colors mixed together. Two centuries later, chemists Robert Bunsen (of gas burner fame) and Gustav Kirchoff developed an analytical technique that involved heating chemical compounds in a flame. When heated to a critical temperature, atoms of different elements become incandescent—emitting light at certain wavelengths, just as wine glasses of different sizes resonate at different pitches when stroked in a glass harmonica. They saw that the spectrum of light emitted by different elements revealed incomplete spectra in which the normal continuous rainbow of colors seemed divided into a number of bright and dark lines, rather like a barcode.

Each bright line is caused by the electrons in each element's atoms jumping and then falling back between higher and lower orbits as they circle the nucleus. This process of falling back is accompanied by the emission of packets, or quanta, of energy, whose frequency depends on the size of the jump. Because the pattern of jumps depends on the unique structure of each element's atoms, it differs for each one, and so every chemical element emits its own distinctive barcode. Throwing common salt (sodium chloride) into a flame, for example, causes it to emit yellow light. That yellow is characteristic of sodium, which is why sodium street lights shine yellow.

Different elements give off unique signatures when heated, but the converse also applies. Light is also *absorbed* preferentially at certain unique wavelengths by atoms of any substance. Thus, when light hits and bounces off any space object, not all of that light escapes. Certain parts of the spectrum are absorbed by the atoms making up the minerals on the surface. As a result, dark lines diagnostic of certain substances also break up the object's reflectance spectrum.

They say that the greatest accolade for a scientist's work is to become "part of the furniture." In the case of the Bunsen burner, the Büchner funnel, or the Liebig condenser, they may mean that literally. Sadly, nobody refers to the petrological microscope as the Sorby microscope. Yet they should, for it was Henry Sorby who first developed the ordinary instrument, used by biologists for example, into one equipped with a polarizing light source and a rotating table on which to examine rocks that have been ground thinly enough for light to pass through them.

Sorby was a remarkably self-contained man, who not only never married but as far as anyone knows never even courted, and whose family had been associated with Sheffield's knife-and-tool-making industries since the seventeenth century. Inheriting wealth from his industrious forebears, he devoted his life to caring for his mother, pursuing scientific projects, and developing science education in his home city, where he founded the precursor to the University of Sheffield. His disappointingly dull personal life might explain why he has only ever had one biographer, Sheffield University librarian Norman Higham, who wrote of his subject in 1963: "If he had any mission, it was to establish to his own satisfaction the microscope as a powerful investigator in whatever science it might be effective."

Sorby's strikingly original mind not only commanded the broad sweep but also observed and quantified natural processes in the minutest detail. His microscopic, quantifying approach seems obvious today; but his work with the petrological microscope was not without its detractors. Sorby himself recalled: "In those early days people laughed at me. They quoted Saussure who had said that it was not a proper thing to examine mountains with microscopes, and ridiculed my action in every way. Most luckily I took no notice of them." It was this single-mindedness, combined with a confident disregard for the opinions of others, which marked Sorby out as a great original.

Sorby had been introduced to meteorites by Manchester University astronomer Robert Philips Greg, whom he seems to have met around 1861, when Sorby consulted Greg about the expenses incurred when the great traveling circus known as the British Association for the Advancement of Science had landed on Manchester. (Sorby was thinking of attracting it to Sheffield). By February 19, 1862, we find the laconic note in Sorby's diary: "Send a/cs of meteor to Greg." In June that year he records, "look over meteoric rocks," and five months later, "Read abt. meteors etc." Greg, a foremost authority on meteors and meteorites at the time, egged Sorby on, writing: "You must go regularly into meteorites, it's of vast importance and most interesting." And it is clear that for both, the collaboration brought new experiences—for Greg, of geology and rocks under the microscope; for Sorby, of astronomy and the wider perspective of cosmic history. Meteorites have a history of causing such academic trespassing, with often happy—but not always painless—consequences.

Sorby had already been the first scientist to examine meteorites in thin section. His Sheffield heritage led him on to studying the iron meteorites, and he swiftly passed on to establishing the metallurgical use of the microscope using reflected light from acid-etched surfaces. Sorby wrote (in hindsight in 1897) that when Kirchoff and Bunsen first exploited the analytical possibilities of spectrum analysis, "I was . . . led from the study of meteorites to invent the spectrum-microscope . . . and to apply that instrument to a very great number of different subjects"—meteorites among them.

Spectrum analysis of sunlight had revealed the elemental composition of the Sun (and hence of the Solar System as a whole) for the first time. Using reflected light, it was now doing the same for planets. The only extraterrestrial material at hand in the 1860s was meteoritic, so after delivering his great paper on meteorites' microscopic structure to the Royal Society, Sorby began studying the structure of metals, and then the spectra of other substances (especially of blood, for forensic applications). Thus the idea of comparing the reflectance spectra of astronomical objects and meteorites had been established.

Greg was an early supporter of the idea that meteorites derived from asteroids, which he had tentatively deduced from the observation that more meteorites fall in northern hemisphere summer (April to September), when the Earth is furthest from the Sun and nearest to the Asteroid Belt. However the most persuasive evidence linking meteorites with asteroids was still waiting to be found—in the comparison of their reflected light.

To find the reflectance spectrum of a meteorite, modern researchers do not use a microscope as Sorby did, but grind the meteorite into powder and bounce light off its compacted surface. The resulting curve is then compared with spectra obtained by analyzing the sunlight reflected from various asteroids, and then looking for the best match. The comparison is not straightforward because asteroid surfaces have been exposed to cosmic rays for millions of years, so their surfaces no longer have precisely the same optical properties as a freshly ground-up meteorite in the laboratory. This so-called space weathering makes asteroids more reflective at longer wavelengths, which astronomers call reddening. However, one look at the next diagram will show that, using this method, meteorite orphans have now been successfully matched to their parent bodies.

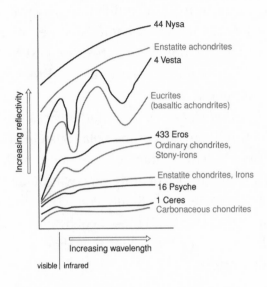

44 Nysa

Enstatite achondrites

4 Vesta

Eucrites
(basaltic achondrites)

433 Eros
Ordinary chondrites,
Stony-irons

Enstatite chondrites, Irons

16 Psyche

1 Ceres
Carbonaceous chondrites

Increasing reflectivity

Increasing wavelength

visible | infrared

Illustration MH

*Reflectance spectra of asteroids compared with
those of certain meteorite types. The offset
between the lines is due to reddening of the
light from asteroids—a feature due to space
weathering by cosmic rays on the surface
materials.*

This surely is one of those moments when you should
allow yourself to be struck by an amazing scientific achievement that seems to have happened by stealth. As a result of
the discoveries linked through this chapter, scientists today
can take a rather undistinguished-looking, grayish rock with
a black fusion crust, formed long before anything that we can
see today on the surface of the Earth even existed, and say
a number of things with reasonable certainty about it. Most
amazingly, they can tell from which small, tumbling, gray,

potato-shaped space-rock, 200 million miles away, it came. In many cases they can tell when it left its home, and how long it has spent wandering, homeless, across the vast tracts of inner space, before finally making earthfall. And they can tell you when that happened too.

Just as Charles Darwin's grandfather Erasmus envisaged in verse, the bright spheres of the Solar System accreted from the chaotic nebula, leaving their leftover debris to drift through the void, while the surface of our planet was made and remade again and again. When they finally come down to us, by contrast almost unchanged in all those aeons, they come as fossils from the dreamtime of the world that they, and they alone, allow us to recapture.

3

THE FALLING SKY

Every great advance in natural knowledge has
involved the absolute rejection of authority.

T. H. HUXLEY

The triumph of the scientific method has taken a long time; but meteoritics as a science began through a set of coincidences—astronomical, social, intellectual and political—in an era of upheaval during the remarkable decades that brought the eighteenth century to a close. During this time of republican ferment in the West, so frequent did meteorite falls become that at times it really must have seemed that the sky was falling. Yet only at that particular time could something that had occurred throughout history (and the reports of the peasants who were generally the witnesses) be seen in a new and productive light. This light was, of course, the light of reason. Before it dawned, however, things were rather different.

During the four years from AD 218 to 222, the mightiest empire of the ancient world found itself officially worshipping a meteorite. This little-known episode in Rome's history ended with its dissolute priest-emperor being hacked to bits and hurled into the Tiber.

The story began with the fall of a blackened, cone-shaped piece of space debris somewhere in the deserts surrounding Emesa—the modern-day Syrian city of Homs. This then appears to have fallen into the hands of a group of nomad king-priests, who settled at Emesa to found a temple to their sun-god *Ilâh hag-Gabal*, or Elagabalus as the Romans came to know and revere him, and of whom the stone was believed to be the embodiment.

A Greek civil servant named Herodian tells us in his writings that the stone was a "sacred image . . . not wrought by human hand . . . worshipped as though sent from heaven; on it there are small projecting pieces and markings . . . which the people would like to believe are a rough picture of the sun." Because of its subsequent entanglements, the Elagabalus stone was depicted on many Roman coins and medals. Most of these seem to corroborate Herodian's descriptions and seem strongly suggestive, to anyone familiar with meteorites, of a so-called oriented specimen.

Oriented meteorites are often pointed, with a distinct nose cone created as material melted away during its descent through the atmosphere. Where material has melted away, pits, grooves and points form, in a process called ablation. Many of these ablation features radiate from the nose cone in a pattern that is often not unlike the rays of the Sun.

How the meteorite became, for a short while, the god of Rome is a tangled tale of political intrigue. In about 203, a boy named Varius Avitus Bassianus was born to the dynasty of Emesa's holy kings, becoming a high priest while still a youth. He was grand-nephew to the mother of Emperor Caracalla, who built the massive public baths in Rome that bear his name, but whose construction was not enough to save him from assassination in 217. The boy's scheming and ambitious grandmother Julia Domna, careless, it seems, of her daughter's reputation, seems to have put it about that the boy was in fact the late Emperor Caracalla's bastard son. This she did just as Caracalla's successor, the praetorian Macrinus, who had exiled the boy's family to Syria in the first place, found himself falling out of favor. The army mutinied; Macrinus was deposed, and the boy, now calling himself Elagabalus after his god, was proclaimed in his place on May 16, 218.

Rome had a new emperor; but alas, at barely fourteen it all rather went to his head. The boy, who had already become very pious, decided to take the meteorite with him to Rome, in a colorful imperial progress that took the best part of a year. By the time he arrived, in the autumn of 219, cracks in the young Elagabalus's sanity had already started to show. Anxious to establish his alien god above the Roman pantheon, Elagabalus built a colonnaded temple to the meteorite close by the Colosseum, known at the time as the Flavian Amphitheater.

Tributes from all the other gods were brought, and the meteorite (male, naturally) was symbolically married off to the two female deities Astarte and Urania, appropriately representing the Moon and the universe. This bizarre ceremony occasioned the richest pomp, as the stone, "set in precious gems, was placed on a chariot drawn by six milk-white horses richly caparisoned"—as Edward Gibbon put it in his *Decline and Fall of the Roman Empire*. Elagabalus walked backward before the chariot, holding the reins, and never taking his eyes from the object of his veneration as gold dust was strewn about his feet.

As if this was not eccentric enough in an emperor, Elagabalus defied Roman convention in many other ways. He put his mother in the Senate; he wore colorful priestly robes and rich jewels; he insisted on performing religious rites personally, and frequently compelled senators and magistrates to join in—which they did, as Gibbon wrote, "with affected zeal and secret indignation." The emperor himself then began wearing makeup and dressing as a woman. Tiring of the fried dormice that the Romans considered a delicacy, Elagabalus developed an interest in fancy cooking. He also had an affair with a handsome blond charioteer named Hierocles, asking surgeons for a sex-change—which they explained was beyond the technology of the time. Strangely anxious that all this might make him seem unmanly to the Roman public, he then raped, married, and divorced in fairly short order a vestal virgin named Julia Aquila Severa. He allocated important political offices to stage actors—something we now do democratically—while finally he is even said to have prostituted himself in taverns.

If only half of these stories were true, it would come as no surprise that the Roman military soon blushed over their choice of emperor, and on March 13, 222, the Praetorian Guard duly put an end to Elagabalus's peculiar reign. The consequences of one meteorite's fall long before in the Syrian desert finally worked through to their conclusion. Elagabalus's mutilated body was dragged through the streets and dumped in the public sewer. Later, the meteorite was quietly returned to Emesa—where it was probably smashed some time in the fourth century, when Christians descended on the cult's temple. They were supplanted in turn by the Muslims, who turned the site into a mosque.

The Elagabalus stone is far from being the only meteorite in history to have become an object of veneration. Indeed, it is possible that one is still worshipped as the *al-Hajar-ul-Aswad*, at the eastern corner of the Kaaba, the large black granite-and-marble cube at the center of the Grand Mosque in Mecca, focus of Muslim prayers the world over. During the Hajj, faithful pilgrims circle the Kaaba seven times, and attempt to kiss the stone (or, failing that, point to it) once during each orbit. It has been much broken during its long and troubled history, and is now mounted in a silver frame. Although evidence is conflicting, the stone is held to have fallen in order to show Adam and Eve where to build their temple—linking heaven and Earth by crossing from one realm to the other. It is possible that veneration of the Black Stone was a much older tradition that was Islamized in the same way that pre-Christian sacred sites and traditions in Europe were colonized by the new faith.

As these tales show, people of the ancient world found little difficulty in believing that stones could rain from heaven. And during the next millennium, the illiterate rural masses of Europe also found little trouble with the idea—particularly as they tended always to be the eyewitnesses. When the book-learned disbelieved their accounts, they too had their reasons. Unlike common folk, the educated were subject to certain precepts—namely the weighty pronouncements of grand philosophers (and, later, physicists), all of whom had long taught that such things were impossible. Heaven and Earth simply did not come together.

The learned proscription against the notion that stones could fall from the sky had begun with Aristotle in the fourth century BC, when the great philosopher wrote his *Meteorologica*. The verbal echo between meteorite and meteorology, the science of our climate, may seem illogical today, but both worlds derive from a Greek root which means "thrown up in the air." Aristotle's book treated everything on Earth and in the air, from stars and comets to auroras, weather, and things that fall from heaven. To Aristotle, fireballs and comets belonged to the sublunary sphere, or lower region of the cosmos, which sat between the Moon and the Earth. Beyond that lay the region of the stars in its unchanging and timeless perfection.

For Aristotle, meteors were exhalations from Earth—a view repeated by the influential Roman natural philosopher Pliny the Elder around AD 75, shortly before meeting his end under the exhalations of Vesuvius in AD 79. Pliny acknowledged that it was possible for stones to fall, as one had done at Aegospotami in Thrace about 467 BC—an event referred to by Aristotle himself. However later commentators, like Roman historian Plutarch, ruled out the idea, saying that such events would compromise the heavens' perfection. Although much of Aristotle's cosmology was demolished in the sixteenth century, his ideas still held sway over sublunary goings-on, including meteorites, long into the eighteenth century, largely thanks to the greatest scientists of the seventeenth century, who upheld the Aristotelian view.

No scientist ever cast as long a shadow as Sir Isaac Newton. Newton, whose mechanics had explained the motion of the heavens, was implacable. Although there was room in his cosmos for "great bodies, Fixed Stars, planets and Comets," there could be no grit in the celestial clockwork. "To make way for the regular and lasting Motions of the planets and Comets, it is necessary to empty the Heavens of all Matter," he wrote in one of his unpublished scientific papers. His contemporary, the brilliant but physically ill-favored Robert Hooke, was of the same view. Although Hooke had observed in experiments that bullets dropped into fine powder produced craters just like those on the Moon, he discounted an impact origin because "it would be difficult to imagine whence these bodies should come."

This combined weight of authority from Aristotle to Newton was hard for learned men to resist. What is more, although it is easy to condemn as snobbery the failure of savants from former times to credit the eyewitness accounts of common people, their failure is much easier to excuse when one remembers all the other things that such folk also found no difficulty in believing: including rains of fish, frogs, flesh, and blood; lizards; sugar-candy; snakes; spiders' webs; and jelly.

By the time of the mid- to late eighteenth century the seed of scientific rationalism, sown in the seventeenth, had taken root. However, savants remained sensitive to charges of credulity. As well as an extraordinary increase in the incidence of meteorite falls, the later years of that century also witnessed a new movement that soon attempted to sweep away the whole rotten edifice of medieval society, with its oppressive monarchies propped up by self-serving superstition dressed up as state religion. And while the rising tide of republican thought was mainly finding expression on the Continent, even in England there existed independent men of spirit whose dedication to truth would overpower whatever instinct they had to cleave to the establishment against the testimony of the poor—always the despised and disbelieved witnesses to the falling sky.

Edward Topham was born in York to Dr. Francis Topham, judge of the Prerogative Court, master of the faculties, a collector of high offices in church and state, and a powerful city figure. For his schooling, Edward went to Eton, where he repaid his father's generosity by heading a notorious revolt over the privileges of school prefects—a rebellion that involved 150 pupils and made the national press. Edward survived fourteen strokes of the rod but remained at Eton. His ambitious father meanwhile turned his attention to soliciting promises from two of his powerful neighbors, the archbishop and dean of York Minster, planning to ensure that various highly prestigious church appointments were conferred upon his young son as they fell vacant. Fortunately for the future development of meteorite science, the parental plot went badly awry.

The dean of York was one Dr. Fountayne, who won his doctoral title thanks to a Latin sermon he delivered at Great St. Mary's Church, Cambridge. What the university presumably didn't know at the time was that the sermon had actually been the work of Laurence Sterne—soon to become a giant of English literature with his wacky masterpiece, *The Life and Opinions of Tristram Shandy, Gentleman*. Sterne already enjoyed many high connections in the diocese of York, including family ties, so a quid pro quo for writing Dr. Fountayne's sermon may not even have been needed to ensure that, in the event, Sterne won preferment in the diocese and not Francis Topham's darling son. The public row that ensued between Topham elder and the dean proved, however, the greater reward for Sterne, who satirized the wrangle in a book that launched his literary career, *A Political Romance* (1759). We are lucky to have this today, for all copies were ordered to be burned under a settlement imposed by the archbishop of Canterbury.

After leaving Eton, Edward attended Trinity College, Cambridge, for four years—where, in a twist of fate worthy of *Tristram Shandy*, he may have unknowingly met its author. Sadly, the brilliant Sterne did not enjoy his prestigious living for very long. His bones, however, which were stolen by body snatchers, enjoyed a brief comeback career as a medical exhibit in the anatomy lectures of professor Charles Collignon of Trinity College. Leaving the university without a degree (almost the rule in those happy times), Topham went on a Grand Tour and then to Edinburgh University, where he studied chemistry, medicine, and anatomy before entering a military career in the Life Guards.

The late eighteenth century brought two great revolutions, in France and America. Topham rose to the occasion. The thin red line was stretched very thin indeed by the Revolutionary War, so to help swell army recruitment, the British government had introduced an act removing various proscriptions against Catholics serving in the military. This inflamed fears among the Protestant majority and led to what became known as the Gordon Riots.

Edward Topham had risen to the rank of Captain by June 2, 1780, when he dispersed a Protestant mob of 60,000 from Parliament Square by leading forty cavalrymen in a successful charge. This deed later earned him an (uncredited) role in Dickens's historical novel *Barnaby Rudge;* but in his own time it won the gratitude of King George III, who with his penchant for snappy one-liners named Topham his "tip-top adjutant."

Topham became the man of the age. Had he lived today, he would probably have employed a publicist and launched himself on reality TV, for his behavior would not be unfamiliar to readers of celebrity magazines. Contemporary cartoons by Gillray and Rowlandson show him setting fashion and having affairs with actresses who had been romantically linked with minor royalty. He fought or seconded in duels, wrote for the theater and, in 1784, met a Mrs. Wells—otherwise known as Cowslip—who bore the retired captain, now styling himself major, four children. Major Topham also founded a newspaper, *The World*, a racy scandal-sheet, and used it to boost Cowslip's career. But the death of their last-born child, a son, led to Topham's final split with Cowslip—an event that had profound consequences for the history of science.

Topham's almost last act as newspaper proprietor was to defend a case in court that established the legal precedent, sacred to journalists to this day, that the dead cannot be libeled. But perhaps the experience was too much even for the king's

tip-top adjutant; once it was all over, he retired to the country, taking his three daughters and, oddly, Cowslip's mother
to a grand property high on the gently rolling chalk of the
Yorkshire Wolds, called rather disingenuously Wold Cottage.

Here he settled into a new life as landowner, farmer, dog
racer, and local magistrate. Not only was he a man publicly
known for obsessive rectitude and truthfulness, he was also
famously incorrupt (an even greater novelty in the eighteenth
century than it is today). A hands-on dispenser of justice, with
a lock-up on his own property, he believed that if a man could
be hanged for theft, it was a scandal that perjury did not carry
a similar penalty. You lied to this man at your peril, and everyone knew it. This lent extraordinary credibility to the testimonies that Topham was soon to collect, following a prodigious
event on his Yorkshire estate.

＊

At half past three on December 13, 1795, the Wold Cottage meteorite dropped into the eighteenth century after 4.5 billion years
in space. It found the major not at home, and it was a ploughman—seventeen-year-old John Shipley—who was the closest
observer, coming within eight yards of ending up as a fatality.

According to sworn testimonies that Topham subsequently
obtained from witnesses, people in the villages around Wold
Cottage heard startling booms "like the report of cannon at
a distance," and felt a ground-shaking thud. Young Shipley
recalled seeing the object when it was about seven yards up,
trailing sparks. A split second later it hit, sending earth flying
up all around him.

George Sawdon, a carpenter, was walking with Topham's
groom, James Watson, about 50 yards away. They ran to the

spot, found Shipley reeling from the shock, and helped the lad to dig the stone out of a crater "about 21 inches deep." The meteorite had passed clean though the thin topsoil, embedding itself into bedrock by about 6 inches. Sawdon noted that it smelled strongly of sulfur and that it weighed "about 56 pounds." It was not only the largest meteorite to fall in Britain (until Christmas Eve 1965, at Barwell); even when considered alongside all the others that peppered Britain, France, and Italy in the closing years of the eighteenth century, it also proved one of the most significant.

Topham first learned of it in "vague accounts" in the provincial and London press, as well as in private letters. Returning immediately to Yorkshire from London, he quickly conducted interrogations. These he reported in a letter dated February 8 to James Boaden, editor of a newspaper called *The Oracle*, with an objectivity that speaks well of his journalistic probity.

For up to three weeks before his return, he wrote, "30 or 40 persons on each day had come to see the STONE which had fallen." He meticulously recounted details from his interviews with witnesses, noting that the stone was "strongly impregnated with sulfur" and had a texture "of gray granite"—a rock completely alien to the Yorkshire Wolds. He noted that it had been a mild hazy day, with no trace of thunder and lightning; yet people had heard detonations which, at the nearby seaport of Bridlington, reminded folk of guns firing at sea. The stone's near-victim recalled "that the clouds opened as it fell, and he thought HEAVEN and EARTH were coming together!"

Topham's letter was republished, with minor changes, in *The Gentleman's Magazine* in 1797; but by that time Topham had already brought the stone to the capital and exhibited it to the public at "No. 2 Piccadilly," opposite the Gloucester Coffee

House, in the summer of 1796. Handbills were printed; an ad in *The Times* dated July 7 tells us that it cost one shilling to get in, and those who paid received copies of the sworn testimonies and an engraving of the stone that may have been by the exhibitor himself, "Mr. Bohner," who was possibly the engraver Thomas Bonner.

The exhibition was later transplanted to Paddington Green; but it is evident, from an indignant letter written by Topham on October 17, that not everybody had been convinced by his sworn affidavits. Nobody knows who exactly it was who questioned the integrity of the exhibit, but historians have speculated that it may have been certain skeptical savants from the Continent, and possibly one Guillaume-François de Luc. Learned opposition to the notion of falling stones remained strong. In his riposte to such doubters, Topham mentions he was "erecting a Pillar on the spot where the Stone fell" to "perpetuate my credulity to posterity." That monument stands to this day, a lonely finger pointing skyward just a few hundred yards from Wold Cottage, surrounded still by the exposed open fields that have changed little since Topham's day.

Much of the research on Topham's story has been carried out by Professor Colin Pillinger of the Open University and his wife, Judith. Pillinger is the charismatic planetary scientist who was the inspiration behind the ill-fated British *Beagle2* mission to Mars, part of the European Space Agency's *Mars Express*. *Beagle2*, due to touch down on the Martian region Isidis Planitia on Christmas Day 2003, disappeared without trace on December 19 in one of the greatest public disappointments of the time. Pillinger, whose Bristolian burr and Tophamesque side-whiskers had become a familiar sight during the build-up to the project, first became interested in Topham while helping to organize a conference on meteorites in 1995 for the 200th anniversary of the Wold Cottage fall. Colin says his whiskers' resemblance to Topham's is pure coincidence; but his understanding of the importance of publicity in gaining support for scientific ideas and endeavors is an even closer one. For just as Pillinger became a household name in the UK (as he drummed endlessly for the support and funding needed for his ambitious £50 million project), what made Wold Cottage a public sensation was the media profile of the tip-top adjutant himself. As Pillinger writes: "he was famous to the point of notoriety, and the mere fact that he lent his name to the event made it a crowd-puller." And as is often the case with publicity, one thing led very quickly to another.

On August 18, 1796 the *London Chronicle* ran an advertisement that read: "This day was published, in 4to. Price 2s. 6d. *remarks concerning stones said to have fallen from the clouds, both in these Days and in ancient Times.* By EDWARD KING Esq." King was a fellow of London's Royal Society and the Society of Antiquaries. The coincidence of his book's appearance with the Wold Cottage fall was no more than that, the author wrote, since he had been thinking about the problem ever since he had heard of an earlier shower of stones at Siena, Italy, in June 1794. In his remarks, King drew comparisons between the texture and composition of the two meteorite falls. King's preferred explanation for falling stones was that they originated where stones are known to come from—the Earth itself. If this were true, they must have been created within, or flung into, the air; and if the latter explanation were true, they must be volcanic in origin. This was also the explanation favored by Sir William Hamilton of Naples, His Majesty's envoy extraordinary and plenipotentiary to the Kingdom of the Two Sicilies.

Despite luxuriating in one of history's greatest ever job titles, Hamilton is best known today for being the cuckold husband of Emma Hamilton, a former prostitute who had been palmed off on the bookish diplomat by his nephew Charles Greville and who (having bewitched the ambassador and married him) went on to become the mistress of Lord Horatio Nelson. This sorry fate has put William Hamilton himself rather in the shade, since as well as being a key figure in the European enlightenment and the Grand Tour, and the discoverer of buried cities at Pompeii and Herculaneum, Hamilton was also the first volcanologist of the modern era—and even had a hand in the birth of scientific meteoritics.

First reports of the Siena meteorite fall reached London's Royal Society from Frederick Augustus Hervey, the wildly flamboyant earl of Bristol and bishop of Derry, an eccentric in a long line, of whom an unknown contemporary wit said, "When God created the human race, he made men, women, and Herveys." The earl-bishop wrote a letter about the shower of stones on Siena to William Hamilton, whom he had known at Westminster School; but for some reason he sent the letter in care of Sir Joseph Banks, at the Royal Society in London. Banks held the powerful position of president of the Society from 1778 until his death—an unequalled forty-one years. During this time Banks, who was principally a botanist, became without question the most influential figure in British science.

According to reports from Siena, at about seven o'clock in the evening of June 16, 1794, a dark cloud had approached the university city out of the northern sky. Rather like the Barwell meteorite, this one too exploded at great height and showered the city in small fragments—one apparently piercing the hat of a small boy and scorching the felt. The fall became one of the most important in history precisely because it took place in front of the Academy, as well as cultured English tourists like Frederick Augustus Hervey taking the Grand Tour. These were no credulous peasants. Stones from the sky became, for the first time, an avowed reality.

In Siena, Abbé Ambrogio Soldani, the professor of mathematics at the university and the "perpetual secretary" of Siena's already century-old Accademia dei Fisiocritici, did precisely what Edward Topham did eighteen months or so later—he collected testimonies. After examining as many fallen fragments as he could find, he published a 288-page illustrated report. Hervey meanwhile had also heard that Vesuvius had erupted just eighteen hours before the Siena fall, so it seemed obvious that he should write to his old friend Hamilton, the acknowledged world expert on the volcano, enclosing a specimen. He reported that local savants in Siena were considering two theories—either the stone had come from Vesuvius, or it had formed from dust in the atmosphere, somehow lithified by lightning. Neither idea was correct, of course; but that they were being seriously considered as explanations marks a leap forward in thinking. For the first time, stones falling from the sky were not being dismissed as the ravings of the uneducated and superstitious masses.

It is significant that Hervey had had personal experience of Vesuvius, and had himself been struck by a volcanic bomb in March 1766 while visiting the volcano (as so many Grand Tourists did) in the company of his old school-chum Hamilton. He had suffered a wound two inches deep, had bled copiously, and been confined to bed for two weeks with a raging fever. This sort of experience leaves psychological scars as well as physical ones; so it was hardly surprising that Hervey favored the Vesuvius theory.

Banks read Hervey's letter and duly forwarded it to Naples, with a disbelieving cover note, suggesting with a wink that old Hervey's fevered imagination had perhaps got the better of him—as, to be fair, it so often did. Perhaps to Banks's surprise, Hamilton was not so skeptical and included Hervey's observations in the first of his many reports to the Royal Society on the state of Vesuvius, which was in a state of frequent eruption at this time.

Those who lived at a greater distance from volcanoes found it easier to perceive two problems with this hypothesis, and Banks liked the look of neither. First, Vesuvius stands 199 miles southeast of Siena, and the fall had come from the northern sky. Second, how could any stone—no matter how it had become airborne in the first place—possibly take eighteen hours to fall to Earth? At the time, Banks probably deferred to Hamilton's judgment. But the arrival in London, later that year, of the Wold Cottage meteorite fueled his skepticism. To Banks and King it appeared to be made of much the same stuff as the Siena stone. Yet it had fallen thousands of miles from Vesuvius. A stone that had apparently flown for eighteen hours was bad enough. But a whole year?

Soldani, meanwhile, had sent another specimen of the Siena fall to a chemist then living in Naples: Guglielmo Thomson. The odd combination of English and Italian names hints at a strange and unfortunate story. William Thomson had studied at Edinburgh and Oxford, was elected a fellow of the Royal Society, and had become well established in Oxford as a physician and a mineralogist. All seemed well, but in September 1790 Thomson suddenly vanished, never to return—reemerging eventually as Guglielmo Thomson of the University of Naples.

These two apparent Thomsons, one with an English "W" and the other with an Italian "G," confused historians for some time, abetted by that lamentable academic habit, which still persists, of listing authors' initials only. Those with a Welsh background might have found the mutation of W to G quite normal, familiar as they are with the Welsh version of William, namely Gwilym—which helps to explain the French Guillaume and of course the Italian Guglielmo. Research by Professor Hugh Torrens of Keele University has revealed that William Thomson did not just leave Oxford—he fled in disgrace, stripped of his degrees. The unfortunate Thompson had been banished, as Oxford University's *Minutes and Register of Convocation* put it, for "sodomy and other unnatural and detestable practices with a servant boy."

Whether Thomson had merely had the misfortune to be caught, or whether, as he attempted to explain in 1790, he had suffered "a most scandalous imputation from an Experiment performed on a man 4 years ago" to do with an anal thermometer, shall perhaps never be known and is perhaps not strictly relevant in any case. Whatever the truth, Thomson traveled abroad for some years, visiting fellow scientists in Paris, Siena, Florence, and Rome before settling in Naples in 1792 in the company of almost sixty ex-pat Englishmen and mutating into Guglielmo.

It was here that he received Soldani's letter and specimen for description and conducted the research that earned him his place in the annals of meteoritics. He noticed the black fusion crust, formed as the meteorite had fallen through the atmosphere, and noted that the interior contained fragments of the unstable mineral pyrites (iron sulfide). He then crushed part of the sample and drew a magnet through the dust—the first time a mineralogical separation had been performed on a meteorite. What he recovered was the strongest evidence yet, though not fully appreciated at the time, that the fallen stone had originated beyond the Earth. He found grains of metallic iron, something never found on Earth, where all iron—a highly reactive metal—is found in combination with other substances, for example, in its oxide or carbonate form.

Soldani included these analyses in his book, incorporating the text of seven letters from Thomson, who mentioned that an (unnamed) friend thought the Sienese meteorite might have fallen from the Moon, perhaps shot from lunar volcanoes. This idea for the origin of fallen stones had first been suggested by William Herschel, the discoverer of Uranus, after imagining that he had witnessed eruptions on the lunar surface in the 1780s. But that mistake hardly matters. The idea that stones could drop from the sky took root in Italy after the Siena fall, and the seeds of this startling idea were sown across Europe, as more and more meteorites fell on its unsuspecting and war-ravaged citizens. Moreover, as the Academy began to take the idea seriously, reports of previous falls—which had either been ignored or actively suppressed for fear of ridicule—began to surface.

With Soldani, King, and Hamilton all unable to resolve a dispute over the origin of these falling stones, and with three hypotheses in contention—Vesuvius, lightning acting on dust, and, from William Herschel, volcanoes on the Moon—Banks, ever the insightful research director, came to a conclusion. If this problem were ever to be resolved, information was needed on the chemical composition of as many of these stones as possible.

Right on cue, yet another specimen then dropped into Banks's hands. At about eight in the evening of December 19, 1798, almost precisely three years after the Wold Cottage fall, a fireball burst in the unclouded sky over a village called Krahut, near Benares—a Hindu holy city of temples on the banks of the Ganges, described once by Mark Twain in *Following the Equator* as "older than history, older than tradition, older even than legend" and looking "twice as old as all of them put together." This piece of typical hyperbole is perhaps not so far from the truth, for Benares claims to be the oldest continuously inhabited city in the world. In 1798, John Lloyd Williams was on the spot and collected reports and specimens which he duly sent to Banks, arriving early the following year. Banks seized the moment and gave his specimens of the Benares, Wold Cottage, and Siena meteorites to a young chemist named Edward Howard for analysis. Howard's work clinched the case for cosmic origin once and for all.

Howard's collection of meteorites for chemical analysis soon began to grow even larger. He obtained one from the mineral collection belonging to the Charles Francis Greville who had palmed Emma Hart off on his good old uncle William. Greville was no field man—he himself had bought the collection from Ignaz von Born, an Austrian mineralogist, the Masonic sponsor of Wolfgang Amadeus Mozart and a possible model for the priestly philosopher-king Sarastro (aka Zoroaster or Zarathustra) in *The Magic Flute*. While it is hard to imagine Sarastro having gambling debts, von Born certainly did; which was probably what forced him to sell his mineral collection to the eager Greville.

During all this scientific activity, the French Revolution was throwing European politics into disarray. While this would in due course have dramatic and positive effects on the progress of meteoritics, by 1800 its main achievement in this regard had been to exile a certain French nobleman and scientist, Comte Louis de Bournon, to the British capital, where in 1807 he was to become one of the thirteen founders of the Geological Society of London. Howard engaged de Bournon to help him, and the Frenchman immediately realized that bulk analysis was not the answer. Like Thomson before him, de Bournon foresaw the need to separate different constituents first and analyze them individually. So, with the help of a lens, he and Howard laboriously picked out, from crushed samples of their stony meteorites, four recognizable components. These they named curious globules, martial pyrites, grains of malleable metal, and earthy matrix.

It was a crucial step, for what Howard and de Bournon found when they subjected these fractions to alkali fusion clinched the case for a cosmic origin. They discovered that the metal grains of the stones and the bodies of the few entirely metallic meteorites they analyzed contained high concentrations (from 9 to 28 percent) of the element nickel. Because nearly all nickel became concentrated in the Earth's core early in our planet's history, it is rare in the crust of the Earth, but remains relatively common in most meteorites. This discovery linked stony and metallic meteorites together as two facets of the same phenomenon and placed that phenomenon firmly in outer space. The Earth and the heavens had been united by chemical bonds.

✳

One saying attributed to Zarathustra by Friedrich Nietzsche in Book 21 (*Suicide*) of his *Thus Spoke Zarathustra* is *"stirb zur rechten Zeit!,"* or "Die at the right time." To scientists, the great prophet might also have said: "Publish at the right time." In 1794, physicist, musician, and inventor Ernst Chladni did just that. The slim volume of 63 pages bore an exceedingly long and rather un-catchy title: *Über den Ursprung der von Pallas gefundenen und anderer Eisenmassen und über einiger damit in Verbindung stehende Naturerscheinungen,* or *On the Origin of the Mass of Iron Found by Pallas and of Other Similar Iron Masses, and on a Few Natural Phenomena Connected Therewith.* This singular book's basic ideas, namely that meteor stones entered the atmosphere from cosmic space, forming fireballs as they fell through the air, flatly contradicted the accepted scientific wisdom of the time. When it appeared simultaneously in Leipzig and Riga, Chladni was widely ridiculed for credulity and for taking liberties with physics. Within only eight years, the accepted wisdom was overturned and meteorite science was established. Chladni's book had landed, as though from thin air, at just the right moment.

Chladni is credited with many great discoveries—for example, the patterns that form when a glass or metal plate covered in loose sand is stroked with a violin bow, a common school experiment to this day. He also invented the glass harmonica, for which Mozart wrote a concerto. This curious musical instrument probably saved him from a pauper's grave in that, never managing to land a permanent job in any institution, its inventor was forced to travel widely from his home in Wittenberg, Germany, giving concerts and demonstrations. Chladni seems to have been a solitary man—he never married—and after dutifully studying law at Leipzig and Wittenberg universities to satisfy his father, when his father died he promptly abandoned his studies and devoted the rest of his days to science.

He led a peripatetic life, hitting the road in a specially made coach containing all his cumbersome apparatuses, instruments, and specimens. He seemed pitiful even to his contemporaries. One letter from the astronomer Heinrich Olbers reads: "Dr. Chladni is here again to give lectures on acoustics and meteorites . . . I have gladly given him his fee, only on the understanding that he will not require me to attend every one of his 12 or 14 lectures. It is truly sad that this . . . worthy man has found no appointment in an institution . . . and at age 67 he must seek his meagre livelihood like this." Although his life was spent alone and on the move, he fathered ideas that lived on, earning him the metaphorical fatherhood not only of meteoritics, but acoustics as well.

Chladni's book established a conceptual framework in which fireballs were nothing to do with the auroras, or lightning, or earthly exhalations. Their speeds were so prodigious that they could only be solid masses, falling under gravity, heated to incandescence by friction with the air. He distinguished shooting stars, which burn up high in the atmosphere, from fireballs, which are large enough to reach close to the ground, before usually exploding, as at Barwell. He then connected the examples of fallen stones to fireballs on the basis of witness testimony that, despite being taken from untutored peasants, he realized was sufficiently consistent between cases to be credible. Chladni the lawyer spoke from experience, just as magistrate Topham did a year or so later.

The motto of the Royal Society is a famous exhortation to scientists to pay no heed to authority—*nullius in verba*, loosely translatable as "take nobody's word for it." True to this essential precept, Chladni's book thumbed its nose at both Newton and Aristotle. No credible reason existed, he wrote, for denying the existence of small bodies in space. To deny them was just as defensible as the contrary. As far as Chladni was concerned, these small bodies could be leftovers from planet formation, or fragments from space collisions—and he was right on both counts. He concluded by calling for the very kind of detailed chemical analysis of meteorites that, five years later in England, finally put meteorites' cosmic origin beyond doubt—at least for savants in most of Europe.

In France, however, the late eighteenth-century flurry of meteorite falls straddled the turmoil that brought in the French Revolution—a political and social upheaval unequalled in historical and philosophical significance since the fall of Rome. Unsurprisingly, the Revolution produced a dramatic change in the way French scientists viewed a phenomenon that had hitherto been witnessed largely by *paysans* who were soon to become *citoyens*. The contrasting stories of two meteorite falls, one preceding and one following the Revolution, serve to show what a fundamental change had overtaken the French intellectual scene.

The French system of trunk roads, the Routes Nationales, began under Napoleon in 1811 as the first rationally designed road system in the post-Roman world. Copied regionally by the lesser Routes Départementales, France soon became crisscrossed with straight roads that paid very little heed to topography. Ancient villages became realigned along the new thoroughfares. Old streets were sidelined. Outside these settlements, the ancient meandering country byways that hugged hillsides and connected neighbor to neighbor at a time when most people lived and died within a few square miles, began to sink into the land, sometimes escaping even to this day the arrival of asphalt.

Southeast of Le Mans, Route Départementale 304 strides confidently out of the little town of Parigné l'Evêque. Heading ultimately for La Chartre-sur-le-Loir, it is only a short step to Le Grand Lucé, where the D304 makes a course adjustment—not far from a house still marked on maps as La Chevalerie. Finding it now is not easy because the ancient roads that serve it are deep and narrow, running between gentle spurs and valleys that today—much as in the late eighteenth century—are covered with a patchwork of chalky fields under the plough and squares of carefully managed forest.

At 4:30 in the afternoon of September 13, 1768, peasants at harvest in the fields near La Chevalerie heard strange sounds like whistling, cannon shots, and the lowing of cattle. Looking up they saw an opaque body fall in an arc into a grassy field. Running to the scene they discovered a triangular stone, half buried, too hot to handle. They ran away in fear; but returning later found it cool enough to examine more closely. Weighing seven and a half pounds, it had a black crust over a gray interior. Abbé Charles Bachelay, a corresponding member of the Académie Royale des Sciences and an enthusiastic amateur mineralogist, sent in a report of the fall, including witness testimonies and a sample. The Academy appointed a committee to investigate, consisting of three fancily named monsieurs: Fougeroux de Bonderoy, Cadet de Gassicourt, and an ambitious young Turk by the name of Antoine-Laurent de Lavoisier.

At this stage in his life, Lavoisier—now universally revered as the founder of modern chemistry—had not made the fatal mistake of marrying the thirteen-year-old Marie-Anne Paulze. Her father was one of a caste of forty noblemen called "tax farmers" who, under a widely hated system, collected taxes on behalf of the French Royal Treasury and paid revenues to the royal purse, taking a cut and sharing any windfalls with the king. The monarchy liked the system because while they lost some of the money, nearly all the hatred stuck to the tax farmers. Lavoisier, already a wealthy man, married into this unpopular business; as a result he was branded a traitor under the Terror, paying the ultimate penalty on the guillotine on May 8, 1794. The judge cut short his final appeal, saying: "The Republic needs neither scientists nor chemists; the course of justice cannot be delayed." (Lavoisier was posthumously exonerated and pardoned eighteen months later.)

Despite this premature death at fifty, Lavoisier completely reformed chemistry by discovering oxygen and hydrogen, by dispelling the erroneous phlogiston theory of combustion and, by showing that matter is neither created nor destroyed in chemical reactions, enunciated the principle of conservation of matter. However he did little to help the emerging science of meteoritics by lending his name to the investigation of the Lucé fall. In Lavoisier's defense, it should be remembered that his name is most usually associated with the report simply as a result of his subsequent fame. Nevertheless, many of the report's erroneous conclusions did find their way into his influential book, *Traité élémentaire de Chimie* (1789), which upheld the idea that meteorites were formed by the action of lightning upon the upper atmosphere, where Lavoisier supposed there existed a "stratum of inflammable fluid" that gave rise both to fiery meteors and auroras.

The report was openly dismissive of the idea of falling stones. "Real scientists," it said, had always regarded the subject as "highly dubious." It then went on to recount the events related by the good abbé, measured the meteorite's density (higher than most siliceous rocks), then described its texture and finally the results of wet and dry chemical analysis. The authors noted the abundance of sulfur as well as metallic grains. They saw the black fusion crust and interpreted it correctly as a glassy coat of rock that had melted. However, they concluded wrongly that its thinness meant that the stone could not have been hot for very long.

This was a false inference; rock is a good insulator, and the hearts of freshly fallen meteorites are often extremely cold, despite having been raised to incandescent temperatures on the outside for many minutes during their fall. However this mistaken assumption led Lavoisier and the others into a greater error. The most likely explanation for momentary melting, they said, was that lightning had struck the ground where a perfectly ordinary sandstone had happened to be rich in iron sulfide. The iron, they speculated, had perhaps attracted the lightning. And the peasants who said they had seen something fall had been . . . mistaken. They were, after all, only peasants.

While France was busy with its Revolution, the sky continued falling. England, Italy, and Germany (through Chladni's book) became alive to new ideas about the origin of thunderstones. Between 1700 and 1750, only nine meteorite falls had been reported, and four of those are now regarded as dubious. Yet in the next four decades twenty-nine falls were reported (six doubtful) and between 1800 and 1809 a further nineteen falls occurred (four doubtful). A change in the flux of meteorites to Earth is of course possible, but is not likely. It is much more likely that, with growing awareness and acceptance by the establishment, more people lost their fear of ridicule and came forward to report falls.

A post-revolutionary meteorite fall in what is now the Département of Orne, Normandy, in the commune of L'Aigle, showed how different the reformed Institut National was from its pre-revolutionary royal predecessor. Fully imbued with the new republic's frenzy for recording, documenting, and centralizing the whole of France, the Institut dispatched one of its brightest young scientists to the scene. No longer would the word of the citizenry be ignored by the new republic's men of science.

L'Aigle was a prodigious fall, showering the area with an estimated 3,000 separate stones at 1:00 p.m. on April 26, 1803. By May 9, a letter had been read at the Institut, and by June 19 an analysis of several specimens communicated. Yet despite this ample coverage, Jean-Baptiste Biot, professor of mathematical physics at the Collège de France, was dispatched by France's interior minister, the scientist Jean-Antoine Chaptal, to conduct a nine-day investigation. Perhaps such exalted orders explain why Biot departed immediately, despite the fact that his wife was about to give birth. By July 4, two days after his son Edouard Constant came into the world, Biot was back in Paris. Fatherly duties must have been light because less than two weeks later he was reading his report to the Institut.

Biot, whose name is immortalized in the mineral biotite, the mica that gives granite its black specks, was said by the French critic and polymath Charles Sainte-Beuve to have been "endowed to the highest degree with all the qualities of curiosity, finesse, penetration, precision, ingenious analysis, method, clarity, in short with all the essential and secondary qualities, bar one, genius." The English mathematician Olinthus Gregory said (in 1821) that he had "never met so strange a compound of vanity, impetuosity, fickleness, and natural partiality."

Nevertheless, meteoriticists take a kinder view of a man whom they place second only to Chladni among the shining stars of the meteoritic pantheon.

Despite events at home, Biot did not skimp his investigation—far from it. He circled in on his ultimate destination, L'Aigle, interviewing everyone in about a dozen farms and hamlets in an attempt to determine how far away the projectile and its attendant booms could be heard. He asked about the height of the fireball when observed, the point of its detonation, and the time. He mapped the area where stones fell, and so drew the very first ever map of a strewn field, defining an elongated ellipse lying northwest to southeast, measuring 6 miles long by 2.5 miles broad.

In massive detail Biot described the tree branches he found broken and the little craters still visible. He noted that the mineralogy of the fragments bore no relation to the geology of the countryside round; nor were any factories to be seen in the neighborhood, of which they could have been some slaggy byproduct. All manner of people testified: old and young, peasants and prelates, men and women, wealthy and poor alike were questioned and their accounts transcribed.

From what they told him, Biot calculated that fireball came in at a declination of 22 degrees—though he could not work out how fast it was traveling. Biot is known to have favored Herschel's lunar hypothesis, and although his report brought him no nearer an answer it now became undeniable that meteorites did indeed fall from the sky over France, just as they did over Italy and England. Savants had listened to the people; evidence was no longer dismissed just because of learned prejudice; and suddenly the historical record seemed to fill with corroborating evidence that few had noticed, or had dismissed.

Biot, a classicist as well as scientist, quoted Lucretius, who wrote in his *De Rerum Natura* (*Of the Nature of Things*):

> *. . . how motion will o'erheat*
> *And set ablaze all objects—verily*
> *A leaden ball, hurling through length of space,*
> *Even melts . . .*

Here was, by a margin of nearly two thousand years, a fine example of a premature theory. It had encapsulated the essential explanation of fireballs, and had been in front of their eyes all that time. Yet because it had been presumed impossible, it had not been seen.

Biot's son Edouard Constant became a sinologist influenced by his father's enthusiasm for oriental astronomy. In the course of his own researches he unearthed the historical writings of Ma Tuan-Lin, which listed astronomical wonders covering the period from AD 687 to 960, straddling the Chou Period and early Sung Dynasty. Believing such events to be the sure harbingers of things to be, Chinese astronomers chronicled them minutely; and Biot *fils* was able to identify 149 meteor shower events from that period alone, all with precise dates. Later he was able to add 272 more from the Sung period (960–1275). The Yuan and Ming Dynasties recorded another 74. This research helped to establish the periodic nature of meteor showers, eventually leading scientists to realize that showers occurred when the Earth crossed, at certain fixed points in its orbit, the dusty trails left behind by comets.

Sixteen centuries after Emperor Elagabalus, the republic of letters had finally established the objective reality of stones from space, and it had begun with listening humbly to the testimony of ordinary folk who, unencumbered with learning, had believed their eyes, just as they had done since time immemorial.

✳

Today, not one but two monuments stand in memory of the meteorite fall at Wold Cottage. One, on the spot where it fell, is Topham's brick obelisk. Field boundaries have changed since 1795, but the young lines of hawthorn, blackthorn, hazel, and briar still follow the grid pattern imposed by the Enclosure Acts that, from 1750—about the time Wold Cottage was built— progressively corralled the open English countryside.

Tawny soil, streaked with creamy white flint gravel, surrounds it. A flaking sandstone plaque, facing the direction from which the meteorite came, bears lettering recently picked out in black at the expense of the Meteoritical Society for the bicentenary of the fall in 1995. Sadly, much of this has already come away under the onslaught of the bleak Wold winter weather. It reads:

Here
On this Spot, Dec. 13th. 1795
Fell from the Atmosphere
AN EXTRAORDINARY STONE
In Breadth 28 inches
In Length 30 inches
Whose Weight was 56 pounds

—

THIS COLUMN
In Memory of it

Was erected by
EDWARD TOPHAM
1799

Long after the hullabaloo died down, Edward Topham, the celebrated hero, fearless editor, and scandalous playwright, died of a relapse after an operation in 1820. His ploughman, John Shipley, who so narrowly escaped death on that momentous day twenty-five years earlier, lived for another nine years—before he too succumbed, apparently rather unexpectedly.

The squat little eleventh-century church of All Saints, Wold Newton, huddles under a tall stand of trees whose black branches, loud with rooks, stirred against the cold, gray sky on the wintry April day when I visited. Just to the right of the path leading to the porch, you can find Shipley's streaked, greenish headstone, which forms, in a sense, the second, more private monument to that prodigious event. It reads:

Erected
TO THE MEMORY OF
JOHN SHIPLEY
WHO DEPARTED THIS LIFE
May 17th 1829
AGED 51 YEARS

All you that do behold this stone
O; think how quickly I was gone
Death does not always warning give
Therefore be careful how you live.

PART TWO

DEMONS

4

TARGET—EARTH

Time and chance happeneth to them all.

ECCLESIASTES 9:11

I have never come anywhere near to seeing a meteorite fall, but I have experienced a collision with a plummeting professor. It was the summer of 1979, and I was a postgraduate student in my second year, studying for a Ph.D on the Swedish island of Gotland. I had arranged two breaks in my fieldwork that year. One was to visit the mainland and meet a number of Swedish geologists in Stockholm and Uppsala; the other was to act as guide to my visiting undergraduate mentor, Professor Derek Ager who, together with his wife, Renée, was touring the Baltic in their Ford Transit campervan, "Goliath."

It was a significant summer for me, because during the mainland tour I had briefly met one Professor Per Thorslund, who four months later would rediscover a specimen sent to him in 1952 and, in a flash of inspiration, realize that it might represent the first example of a fossilized meteorite that had fallen to Earth over 400 million years ago. That discovery would lead eventually to a new understanding of how meteorite bombardments of Earth might actually be good for life on our planet. The other significant event was that it finally changed my relationship with my teacher Derek Ager from acolyte to colleague and friend.

Although neither I nor anyone else could have known it at the time, Thorslund's discovery in the darker recesses of his collections would pave the way for a more general appreciation of yet another idea that Ager, long an advocate of the importance of "catastrophes" in geological history, had been cooking up in his capacious brain. Geologists would soon be forced to come to terms with the then unpopular idea that massive meteorite impacts had affected our planet in the past. Indeed, one such may well have been responsible for the end-Cretaceous extinction. That revolution would happen in the 1980s. Not until the twenty-first century would Earth scientists move beyond this toward accepting Ager's other big idea. Mass extinctions, he had come to believe, could in fact *never* be single-cause events. This notion is still being hotly fought over now.

✳

Shortly before lunch one day, as Ager and I were clambering down one of Gotland's towering coastal cliffs, I heard a faint but growing cry. Looking up, I saw my old professor falling through the air toward me, wearing an expression of extreme apprehension. Events immediately after our collision are a little confused; but we ended up hanging upside down together in a bush, laughing. Our intellectual collision, which had begun some years before we had even met as student and professor, had always possessed a similar quality of alarming but happy catastrophe.

Because it influenced the rest of my life, my first contact with Derek Ager carries with it the compelling illusion of having been pre-ordained. This illustrates one of the most difficult mental efforts that any historian has to make—namely that just because things happened the way they did does not mean they had to happen that way. Our mistake is to assume that because an event turns out to be important, it cannot also have been accidental.

There is a lesson for science here. Pre-scientific attempts to derive meaning from meteorite falls and fireballs always began with a very similar assumption—namely that such prodigies must be deliberate and must therefore be imbued with meaning as signs, for good or evil, destined by providence to warn us of what awaits. Science, on the other hand, has had the effect of turning meteorites into the complete opposite—not only making them the relics of the unimaginably distant past, but turning them into a perfect metaphor for the chance that, as the preacher in Ecclesiastes tells us (in the accidentally portentous verse number 9:11), "happeneth to them all."

In fact nothing but blind chance blew me one day, as a schoolboy, into Swansea Central Lending Library, where I first picked up a copy of Derek Ager's textbook *Principles of Palaeoecology*. This was the collision that really mattered to me. Palaeoecology is the study of fossil organisms as living things—divining their life habits from their physical structure and, by combining paleontological information with what their entombing sediments tell us about their environment, reconstructing vanished living communities. "Fossils were once animals and plants that lived and breathed, fed and bred, moved and died," the book told me. I knew almost immediately that this was what I wanted to do—and eventually came to do on Gotland.

As a boy scientist, my main interest had been the study of ecology, and I had spent long hours with a notebook and quadrat, mapping out different plant species growing wild on the Pennant Hills behind the house where I grew up and where my parents had courted during the war. A subject that offered the chance to combine that old interest with a new one that I had lately acquired, namely geology, seemed irresistible; but my reasons for gravitating toward this book's author were more than just scientific. This textbook—uniquely in my experience then, and I have not seen many other examples since—spoke to me personally. Its writing was elegant and amusing and even contained what, in that context at least, qualified as jokes. I felt the presence of a personality: discursive, irreverent, slyly controversial and yet self-deprecating, as if with typical English reserve he was embarrassed to be rocking the boat, and used laughter to soften the impact of his more disconcerting insights.

As luck would have it, and by yet another random chance, I soon discovered that Ager had quite recently moved from Imperial College, London, where the book's title page said he was based. He had lately moved west, and taken the chair in geology at Swansea University—then one of the UK's most dynamic Earth science departments (now sadly long gone). I didn't even need to leave home. The mountain, for once, had come to Mohammed.

Ager was a reluctant revolutionary. His palaeoecology book had been the first in English (the only previous one had been in Russian), and although he was the UK's foremost exponent of this new subject he rather disapproved of anyone who dared study it without first serving time, as he had, in the engine rooms of formal, taxonomic paleontology. You had to earn the right to think of fossils as living things by first doing an apprenticeship, describing and naming species yourself. Then there was continental drift. Ager was, of course, fully converted to plate tectonics by the time I first met him in the early 1970s; but he had long held out against drift (rather unusually for a British soft-rock geologist of the time)—and had said so in print as late as 1961, in his beginners' book, *Introducing Geology.*

But plate tectonics were someone else's business, and Ager was content to accept the result of that particular revolution once it was achieved. The revolution that he spearheaded himself was in my view no less radical—namely, the rehabilitation of the "rare event" in Earth history. This move constituted a dramatic shift in the way that geologists interpret the past from the evidence in the rocks and importantly paved the way for them to see the Earth within a cosmic environment that could influence terrestrial events. Earth could legitimately be seen as subject to, for example, catastrophic meteorite impact. Such catastrophes might then be legitimately considered as possible causes, say, of mass extinctions; though this idea, when it entered the scientific scene so forcibly in 1980, was yet another theory with which the reluctant revolutionary found difficulty—albeit for reasons other than distaste for the then widely and old-fashioned manifestation of catastrophism.

It is not hard to understand why catastrophic events were so frowned on by conventional Earth science at that time. Before the birth of what may be termed scientific geology at the very end of the eighteenth century, when scientific meteoritics also began, geological thinking had been dominated by grand theories of the Earth that very often invoked global catastrophes, great deluges and massive, sudden upheavals. While this catastrophist approach continued, geology could hardly lay claim to being scientific because there was no natural check on the geological imagination. All ideas, to be scientific, have to be testable against reality. In chemistry, theories can be disproved or allowed to pass on to their next test by experiment on a laboratory bench. But where was the laboratory of the geologist? Until they could find a way of testing their speculations about the past, there was no real science in what geologists did.

However, if you assume that the past was not really so different from the present, and that the natural processes operating today have always operated, then the laboratory of the geologist becomes the world around us. Rocks can be interpreted with reference to processes that can be seen going on anywhere, in seas and rivers, deserts, volcanoes, and even the deep Earth.

This idea—called Uniformitarianism, which sounds like a religious cult, but is more a doctrine than a single idea—is traditionally traced to the pioneering Scottish geologist James Hutton. Hutton, rather like Socrates, is known mainly from what others have written about him—in Hutton's case, because his writing style was thought to be impenetrable. As a result almost nobody has ever read his *Theory of the Earth*—and those who have interpreted him for later generations have effectively discouraged others from trying, possibly one suspects so that their readers will not discover how they, allegedly Hutton's disciples, have subtly altered his original message for their own purposes.

As well as being a uniformitarian who pleaded for a very ancient Earth, Hutton was also a volcanist and recognized the importance of the Earth's internal heat in driving geological processes. He proved from his field observations on Salisbury Crags, which towered over his (now demolished) Edinburgh home, that some rocks had once been molten and had later solidified. Hutton's view of volcanoes (in his own words, written in his *Abstract,* 1785) were that they were "the proper discharges of a superfluous or redundant power; not as things accidental in the course of nature, but as useful for the safety of mankind, and as forming a natural ingredient in the constitution of the globe." In Hutton's view, rare violent events, natural catastrophes, deserved their place within his grand overall scheme.

Without doubt the greatest disciple, proselytizer, and not-so-subtle twister of Hutton's ideas was Charles Lyell, in his *Principles of Geology* (first edition, 1830–33), probably the most influential Earth science book of all time. The rest of its full title gives the gist: *Being an Attempt to Explain the Former Changes of the Earth's Surface, by Reference to Causes Now in Operation.* To use yet another famous paraphrase, "the present is the key to the past." Charles Lyell's version of uniformitarianism, which held sway for over a century, was very different from Hutton's original.

Uniformitarianism encompasses two major aspects—uniformity of process and uniformity of rate. Lyell, reacting (perhaps over-reacting) against the bad old ways of unconstrained catastrophist thinking, imposed a particularly severe form of uniformitarianism on geologists, one that cast a shadow from which we did not fully emerge until the late 1970s, at least in part thanks to Derek Ager. In Lyell's worldview, the only permissible geological agents were "causes now in operation," erosion by waves or winds or rivers, slow deposition in seas and lakes. Inherent in this is the understanding of what "now" means. In Lyell's view, "now" was restricted to the timescale of human civilization—which we realize today is actually vanishingly small in relation to the history of the Earth. Moreover, the rates at which those everyday, observable processes were allowed to operate under Lyell's stern gaze were maximally slow—and their effects infinitely gradual. In this view of the geological past, massive results—such as the accumulation of great thicknesses of sediment, for example—were attainable only by tiny but inexorable degrees, achieving their huge work over vast and rather unexciting tracts of time.

Although in Lyell's world, as in Hutton's, sudden things did happen—the eruption of Vesuvius, for example—the gradual actions of erosion and re-deposition would soon eliminate all but the faintest traces of such unrepresentative paroxysms. The dominant signature in the record of the rocks would not be these out-of-the-ordinary events, but what Ager would caricature as the "gentle rain from heaven." Anthony Trollope's novels are peppered with dogged, persistent characters—usually ill-favored suitors or doughty and sometimes dowdy daughters—who struggle long and hard to achieve their dreams. Trollope's favorite image for these admirable souls' wars of attrition on a recalcitrant world is: "like the drop of water that wears away the stone." I often wonder whether Trollope read Lyell.

This difference between Lyell and Hutton did not go unnoticed when the first volume of his *Principles* appeared in 1830. The English philosopher of science William Whewell (the man who coined the word "scientist" and hundreds of others) reviewed the book in January 1831 in *The British Critic*. He noted: "Hutton . . . did not disdain to call in something more than the common volcanic eruptions which we read of in newspapers from time to time. He was content to have a period of paroxysmal action—an extraordinary convulsion in the bowels of the earth . . . Mr. Lyell . . . requires no paroxysms." Reviewing the second volume in 1832 in the *Quarterly Review*, Whewell not only coined the terms "uniformitarian" and "catastrophist," but also made the point that only properly came to be understood 150 years later, when Ager reminded the world of it by rehabilitating the importance of the rare event within the compass of uniformity. Whewell wrote: "It seems to us rather rash to suppose, as the uniformitarian does, that the information we at present possess . . . is sufficient to enable us to . . . include all that is past under the categories of the present. Limited as our knowledge is in time, in space, in kind, it would be very wonderful if it should have suggested to us all the laws and causes by which . . . natural history . . . viewed on the largest scale, is influenced." In other words, we simply haven't been here long enough to have witnessed everything the Earth can do. He was the first of many to think this thought. Thomas Huxley, in his presidential address to the Geological Society in 1869, said: "It is very conceivable that catastrophes may be part and parcel of uniformity," and went on to point out how, while a clock may tick (Lyellian uniformitarianism) it may also, occasionally, strike (catastrophe). This idea did not go away—it hung around, like an unresolved specimen hidden away in the geological subconscious, until Ager dusted it off.

By the time I had become Ager's student, his worldwide field experience was telling him that Lyell's view that nothing large and lasting was ever achieved quickly in nature, was seriously flawed. He realized that one rare storm—a hurricane, for example—could transport and deposit more sediment in a single night than years of more ordinary conditions. Ager liked his catchphrases, and often quoted Thomas Mordaunt, saying that the sedimentary record was biased—but not, as Lyell had maintained, in favor of the age without a name. Rather, it reflected a few widely separated but crowded hours of glorious life.

Another of Derek's aphorisms compared the history of the Earth to the life of a soldier—"long periods of boredom and short periods of terror." And in the geological record, those short periods of terror were what mattered, in the sense of being most likely to give rise to something that would be preserved in the sedimentary record. (This saying of his has skimmed like a stone across the pond of science writing, being quoted first by Stephen Jay Gould in *The Panda's Thumb* and more recently by Bill Bryson in his *A Short History of Nearly Everything*.)

But that is the sort of writing talent you need to get heretical ideas across, and by the early 1970s Ager realized he needed to write an "ideas book"—a manifesto that he could nail to the door of the uniformitarian cathedral and like Luther say, "Here I stand; God help me I can do no other." This book could have been some mighty treatise, but thankfully Ager had much more of Voltaire than Karl Marx about him, and the result was a slim volume called *The Nature of the Stratigraphical Record* (1973).

The book hit the shops just as I became an undergraduate. He wrote it, he said in its introduction, because he had to: because the ideas it contained had been "fermenting in my brain for years, and I had to write them down before I became completely intoxicated." Its appearance—aided by its brevity and readability—encouraged others, and marked the start of Ager's revolution. The rare event, the catastrophe, had been rehabilitated. Although rare to our human eyes, in Earth's perspective they were almost everyday events. The new approach to Earth history was soon christened neocatastrophism. Occasional catastrophes became part and parcel of the ordinary operation of our planet. On the Earth's vast timescale, they were, in fact, perfectly uniformitarian after all.

Ager might have added "or meteorite strike" to an opening list of natural catastrophic agents when he wrote: "The hurricane, the flood or the tsunami may do more in an hour or a day than the ordinary processes of nature have achieved in a thousand years." But in 1973, when he wrote it, that could have been a step too far. You must search quite hard to find any mention of extraterrestrial influences in *The Nature of the Stratigraphical Record*. No doubt Derek felt he was risking enough simply by re-introducing the word "catastrophe" into mainstream geology, without straying into the fevered realm of cosmic influences.

It is hard to squeeze our feet back into the pinching intellectual shoes of the early 1970s. It was a very different world. For example, at that time in Britain it was not normal for students—or indeed anyone—to bathe or shower more than once a week, and those few who did caused comment, which is why I remember it. Nor had anybody, unless they had visited America, ever seen a McDonald's restaurant. The very first one opened for burgers and fries in 1974 on Powis Street, Woolwich. For the first time, British people saw what a really clean restaurant was like—and mistook it for glamor.

As incomprehensible as that may sound, it is surely no more remarkable that at that time paleontologists were reluctant to talk about mass extinctions, or even acknowledge them, whereas nowadays it is difficult to get them to talk about anything else. In true Lyellian fashion, they had grown used to rationalizing them out of existence. *Natura non facit saltum*—nature does not make jumps, they said. Perhaps these extinctions only appeared sudden because of slow deposition. Perhaps they were merely quickenings of standard processes. Or perhaps they were artifacts—created by paleontologists' tendency to work on the rocks of a certain age and place, and not to stray into other eras and areas—hence their species stopped at self-imposed boundaries. The way Ager wrote of mass extinctions, with the words "There is no point in denying them," is a clear indication of a then contemporary tendency to do precisely that.

Ager made one reference to meteorite strikes in his book, on the authority of a very distinguished paleontologist, Digby McLaren, who had mentioned them as a potential agent of mass extinction in a presidential address to the Paleontological Society of America in 1970. In another publication, Ager also refers to the earliest of the "modern" geologists to suggest meteorite strikes in this context—namely, Max De Laubenfels of Oregon State College, Corvallis, who published the idea in 1956, having kept it under his hat since 1937. Citing such ideas in the 1970s took a scientist very close to the edge, and for some, beyond it.

Appeals to astronomical agencies like comets, meteorites, supernovas or clouds of cosmic dust were all highly distasteful to geologists at that time (and to nearly all legitimate scientists). Such ideas had already become the mainstay of the pseudoscientific popular fringe—with whose outpourings bookshelves of the 1970s were groaning almost as loudly as the geologists or archaeologists who occasionally stooped to read them. Swiss author Erich von Daniken invoked alien visitations to explain all manner of ancient marvels, principally in *Chariots of the Gods*, published in 1968. The catastrophist works of psychologist-turned-prophet Immanuel Velikovsky (originally published in the 1950s) achieved newfound popularity with their exciting accounts of planets wandering about the Solar System with wanton freedom (and catastrophic consequences) even within the span of human history. If the risk of being compared with them was not enough to discourage any legitimate scientist from considering cosmic influences, there was another factor. Impacts violated geologists' parsimonious instincts.

One of science's central assumptions, often known by the shorthand name of Occam's razor, states that the simplest plausible explanation for anything should always be preferred. For this reason, looking outside the Earth for explanations of events in Earth history didn't seem proper at that time. Scientifically, such an appeal seemed trite and unsatisfying, like a *deus ex machina*. Moreover, it was straying into another discipline, at a time when such boundaries, which now seem arbitrary, felt like iron curtains. Earth scientists today have become completely accustomed to thinking of the living Earth as a single system, which can only be properly studied if we look at it through a seamless amalgam of geology, biology, oceanography, meteorology, and astronomy. We have all seen the famous *Earthrise* picture, taken by Apollo 8 astronaut William Anders on Christmas Eve 1968, until it has become a cliché; nevertheless, its clear message has sunk in.

So although Ager, reluctant revolutionary as ever, remained wary of going too far, others—younger, and with far less credibility on the line—were not. Ager drew toward him research and post-doctoral students eager to work in this emerging field of neocatastrophism; and I well remember the frisson of forbidden delight that passed through the Small Lecture Theater one evening when one of these, Richard Hodgkinson, who arrived sprinkled with the stardust of having previously worked at NASA, delivered a seminar about impact craters on Earth.

Barely five years later, and by then a researcher myself at a different university, I found myself listening to much wilder ideas without turning a hair—including one that connected episodes of extinction to bombardments, regularly spaced at 26 million years apart, generated by objects from the distant Oort Cloud, which, the story went, were now and then catapulted out of their peaceful existence by the gravitational perturbations of a mysterious and undiscovered companion star to the Sun called Nemesis. Then they became comets bound for the inner Solar System and a rendezvous with Earth. There will be more about this idea later, but suffice it to note here that it was being given a serious hearing even then.

The revolution had triumphed. Lyellian gradualism, which had overstayed its welcome and prevailed for an age without name, had been swept away in a short and slightly terrifying period of glorious life. Meteorite impacts and the scars they left behind were accepted by geology at last. How long that struggle had taken can be judged from the story of the Earth's most famous visible impact crater—the first ever to be recognized as such—and the two contrasting men of science who were the first to become embroiled in trying to understand how it had been created. For without this story, none of the later understanding we have reached about impacts and their importance in the history of Earth and life could ever have come about.

✳

No one knows who exactly discovered the Earth's most well-known impact structure still visible to the untrained naked eye, in Arizona. In fact, the plethora of names by which it has been known—Coon Butte, Coon Mountain, Crater Mountain, Barringer Crater, Meteor Crater, and now officially Barringer Meteor Crater—testify to the fact that for a very long time nobody quite knew what to make of it. Native Americans naturally were there first, and the site is rich in artifacts sig-nifying that they held it in some reverence. However, what caused confusion right from the start was the fact that it first came to scientific attention not in its own right but because of its association with pieces of nickel-rich iron strewn over a five-mile radius—a circle that also included, to the west of the crater, a generally dry, meandering gulch known as *Cañon Diablo*—Devil's Canyon.

The first iron mass was discovered by shepherds grazing their charges in 1886. Next on the scene were prospectors who, thinking they had stumbled on a vein of iron ore, sent pieces off for analysis. One of those who received samples was Arthur Foote, a Philadelphia mineral dealer. He knew at once that the samples were meteoritic and immediately made plans to visit the site, which he did a few weeks later. Foote gathered further frag-ments of the meteorite that became known as the Cañon Diablo Iron, and published his findings in the 1891 *Proceedings* of the American Association for the Advancement of Science (AAAS).

The paper created a stir on account of the microscopic dia-monds that Foote reported finding when a specimen of the meteorite proved to be harder than the emery wheel being used to polish it. These diamonds had been created from carbon within the meteorite under the shock of impact; but at that time the special effects of shock metamorphism were poorly under-stood. The diamonds misled scientists into assuming that the parent body of the Cañon Diablo Iron had been a planet large enough to create diamond-forming pressures in its interior—which would have made it as large as, or larger than, the Moon. This created a dilemma, since, as we have seen, all the asteroids in the Asteroid Belt today would, if assembled in one object, make up a body with less than 3 percent of the Moon's mass.

Only once in his paper on the Cañon Diablo Meteorite did Foote address the small matter of the gigantic impact crater, .7 mile across, 197 yards deep, whose rim rises 55 yards above the surrounding plain, which was lying plumb in the center of the Cañon Diablo strewn field. He believed the crater was volcanic, and presumably that its association with meteorite fragments was coincidental.

This may seem amazing today, though interestingly the same thing has recently happened in reverse. Planetary scientists interpreting new photographs of the surface of Mercury discovered a crater surrounded by a distinctive radial structure known as The Spider. This, to a geologist's eyes, almost certainly represents a collapsed volcano, which explains both the crater and the radiating lines. Nevertheless the crater at the center of these radial valleys has been officially interpreted as the result of a coincidental meteorite impact. The jury is still out, but the dispute illustrates the power of prevailing mindsets to alter the perceived meaning of objects in science, just as in any other field of endeavor.

The Meteor Crater's first non-Native name—Coon Butte—reveals that this huge hole does not present itself as such to a person approaching on foot or horseback. To those eyes, the crater's raised rim makes it look like a mountain. As you drive toward it out of Flagstaff the road drops through six or seven hundred yards as the land becomes progressively more barren until only sagebrush grows on the flat, hummocky desert plain. After 40 miles on Interstate 40 you come to exit 233, where a sign invites you to the National Monument visitor center on the crater's north rim.

There you soon become aware, as you thread through the gift shop and past the talking museum exhibits, that Barringer

Meteor Crater is owned and run by the Barringer Crater Company, Decatur, Georgia, and operated by Meteor Crater Enterprises of Flagstaff. As the historical exhibits explain, the Barringer family has been associated with the crater since 1902, when visionary geologist and mining entrepreneur Daniel Barringer first heard of it.

Foote's 1891 paper was read at an AAAS meeting in whose audience sat Grove Gilbert, chief geologist of the United States Geological Survey (USGS) and the most influential rock-tapper in the land. He was interested in Foote's paper because of his involvement in the then current controversy over the craters of the Moon, and whether they were caused by impacts or volcanoes. Gilbert had been using photographs of the Moon to try to prove that impacts had been responsible. He had long believed that the Earth had formed out of the protoplanetary nebula by the accretion of smaller "planetesimals." So after Foote had finished speaking and comment was invited from the audience, Gilbert got to his feet and proposed that one such late-falling asteroidal body had made the crater at Coon Mountain, and that this impactor might even now lie lodged beneath its floor.

Gilbert asked one of his geologists, Willard Johnson, to write a report. Johnson drew attention to the fact that strata in the crater walls were bent backward and meteorite fragments lay all about, and initially wrote that "a big fellow had made the hole." He noted that Gilbert's meteorite idea had also occurred to local prospectors eager to find a large mass of celestial iron—though no compass-needle deflection had been noticed anywhere around the crater to support the idea of a large buried mass. Gilbert himself—clearly excited, writing to a friend that he was "going to hunt a star"—spent several weeks at the crater in October, creating a detailed map of its topography and searching for any magnetic variation that might be due to a buried iron meteorite.

In common with all other U.S. Earth scientists of his era, Gilbert was an adherent of an approach known as Multiple Working Hypotheses, a distinctively American way of doing science that took its cue from the egalitarian principles of the fledgling republic. U.S. geologists had an enormous task to document a new and largely unexplored continent. Acutely aware that their evidence was no less valuable than evidence gained elsewhere, they rebelled against the notion that their task was to follow the coat tails of Old World science. The republic enshrined principles that aspired to better ways of doing things. Science was not immune from this reforming zeal, and in place of the grand theories of old, where savants had devised all-encompassing systems of the world to which evidence was bent to fit, American science would be truly egalitarian.

Building on an emerging Anglo-Saxon model of facts first, theories later, U.S. science had an ideology that prided itself on its lack of ideology. Democratic through and through, every theory would be given equal access to light and sustenance. None would be unfairly favored—there would be no first sons. Only after all lines of evidence had been considered and weighed in the balance would one theory emerge as having grown straightest and most true. In a way, it was a little like Sherlock Holmes's approach, whereby after eliminating the impossible, whatever remains—however improbable—must be the truth. Gilbert therefore adopted a companion for his first-born theory—the idea that the crater had been the result of a massive volcanic eruption involving steam at great depth. He then ran the two ideas against one another, to see which one came out on top.

Gilbert set out three tests that his impact theory must pass if it were to be adopted. The first was habeas corpus—a meteorite mass must lie somewhere beneath the crater. This in turn must mean that, if you could calculate the volume of the upthrown crater rim material and compare it with the volume of the crater, there should be material left over equivalent to the volume of the impactor. Third, though not crucially, the crater should probably be elliptical, because most meteorites will not strike vertically. Sadly, all three of these tests were based on misconceptions and led the poor methodical Gilbert completely up the wrong path.

Returning to Washington, Gilbert had his samples of rock and meteorite chemically analyzed, and began doing his laborious sums. Working from his own topographic map, he showed that the material ejected from the crater came to 82 million cubic yards—alas for the meteorite theory, almost exactly the same as the volume of the hole. Moreover, said hole was more round than elliptical, and displayed no magnetic anomalies at all. (The similarity in volume between crater bowl and uplifted rim, which Gilbert found so convincing, was in fact a complete fluke. The rim has been eroded by 50,000 years of exposure, and so bears no precise relation to the volume of the crater, which itself has been made considerably shallower by the accumulated sediments of a lake that once occupied it during wetter times than prevail today.)

It must have been a bitter disappointment; but true to his principles, Gilbert relinquished his impact theory in the face of these results—and in 1896 published his official conclusion. The theory of a buried star could not be supported. Not for the first (or last) time, a first-rate scientist had ignored his instincts and reasoned himself into error with a convincing mix of heavy effort, false premises, and mathematics. Gilbert concluded that the crater was volcanic. The meteorite fragments strewn around it must be coincidental. This was an untidy hypothesis, but the facts seemed to demand it—and Occam's razor be damned.

Gilbert was clearly not satisfied, and wrote presciently of his hope that one day new knowledge of cratering processes might rescue his original idea. Meanwhile, he did his best to live with it, using the example of how he had methodically rejected his initial, dramatic hypothesis in favor of a much duller one, in a series of popular lectures about how science works. But Gilbert's was not the only game in town.

✳

Nobody becomes a mining entrepreneur if he is easily put off by the misgivings of academics, and Philadelphia lawyer, geologist and mining engineer Daniel Barringer was a man who knew a good bet, even if the odds were long. Barringer already partly owned a successful silver mine in the state at Pearce. As a geologist he was aware that meteorite irons were much richer than any earthly ore. What is more, he knew they were uniquely rich in nickel—a rare and expensive metal, for which America had no domestic source. So, having heard about the crater in 1902, apparently dropping his cigar in surprise, in 1903—without going there himself, lest his presence put up the price—he staked a claim.

This claim was signed by two of Barringer's hunting buddies. One was Owen Wister, author of the first cowboy novel, *The Virginian* (1902). The other was the dedicatee of Wister's novel—twenty-sixth president of the United States, Theodore Roosevelt. The claim was made under the name of the Standard Iron Company, which was duly granted a mining patent to explore for and recover metals. The following year, drilling began in the center of the crater, the reasoning being that because the crater was circular, the meteorite had struck the ground perpendicularly—as we shall see, a common-sense view that has been since disproved by ballistic science. (Barringer was very hot on common sense, which he frequently invoked against scientists whose highfalutin theories were unfavorable to his scientific or commercial case.) Five holes were sunk. They found nothing.

Undaunted, Barringer and his physicist associate, Benjamin Tilghman, published a further defense of the meteorite hypothesis. Barringer in particular lost no opportunity to trash Gilbert's conclusions, citing several criteria that, unlike Gilbert's "tests," would be acceptable even today as evidence for impact—namely, overturned and uplifted rim strata, pulverized rock, and nickel oxide mixed in with sediment. For almost thirty years the controversy raged on, Barringer doing whatever he could to fight off his scientific opponents and bolster his stockholders' confidence. There was the finely pulverized silica. Large quantities of iron oxide and traces of nickel were found mixed with sediment and spread symmetrically about the crater. The strata ejected from the hole had fallen back in reverse order, youngest at the bottom. Moreover, if the crater were volcanic, as Gilbert concluded, where were the volcanic rocks? Gilbert's alternative hypothesis—which invoked the explosion of groundwaters under the influence of rising magma—was just as wild as the meteorite hypothesis. Where in the world was the model for such a cryptovolcano, a volcano with all traces of volcanicity removed?

Barringer was a lusty, powerful, and eternally optimistic man with no time for polite niceties, using words like "absurd" about rival theories and "blind" and "demented" about their originators. Barringer's criticisms of Gilbert, coming as they did from an outsider to the scientific establishment, went down particularly badly. Gilbert was the closest thing to a secular saint ever to have occupied the post of chief geologist at the USGS, and attacking him was tantamount to goosing the Blessed Virgin.

Meanwhile as Gilbert and the USGS held their peace (Gilbert, unlike Barringer, hated public rows), even Barringer's apparent scientific allies were starting to turn into commercial enemies. George Perkins Merrill, curator of geology at the Smithsonian Institution, was an eager believer in the meteoritic origin of the crater. He conducted his own survey in 1907, noting the extreme contortion and shattering in the crater walls, and the abundant fine glassy dust, derived from the sandstones, which, he rightly thought, looked as though it had been formed by "a sharp and tremendously powerful blow." So far, so good for Barringer and co. But then, disaster. Of the meteorite itself, Merrill concluded that its size might have been exaggerated. (It had and would continue to be for some years.) Worse still for Barringer's backers, Merrill suggested that as well as being uncommercially small to begin with, the main mass might have been vaporized during impact.

Barringer could not be expected to react well to the idea that his prize had literally vanished into thin air, and he didn't. But to rescue his meteorite from the vapors, Barringer developed the idea that the impactor had been a swarm of several, whose total weight he eventually suggested may have totaled 10 million tons—much of which he fancied might now be lodged under the south rim. This was where subsequent operations concentrated.

Into the 1920s, Barringer refinanced operations through a series of tie-ups with other companies, and by forming new ones. One last attempt to locate the mass began in 1927 but they found only water—so much of it, in fact, that drilling was stopped. A report commissioned in 1929 by Quincy Shaw, the president of the Meteor Crater Exploration and Mining Company, agreed with Merrill. Either the meteorite was uneconomically small, or had vaporized. Shaw closed down drilling. A few months later, on November 30, Daniel Barringer died of a massive heart attack.

As the century endured a second world war, much more came to be understood about the physics of explosion crater formation—of how impacting objects can be utterly destroyed, and how they leave circular depressions no matter at what angles they may hit. "Common sense," always Barringer's lodestone, was not the infallible guide he thought it was, as is so often the case in science. Depending on its size and velocity, a meteorite may flash to vapor as its immense kinetic energy is converted to heat. At the instant of impact, irrespective of trajectory, the flash to gas creates a symmetrical, round (or, in the case of Barringer Crater, rather square) crater.

The modern mass estimate for the Cañon Diablo Iron, about 300,000 tons, is a fraction of Barringer's 1914 estimate of "10 million tons." If Barringer had been right about that, at contemporary prices for ores as rich as Cañon Diablo it would have been a prize indeed—worth about $1.3 billion. That still sounds a respectable sum; and in 1914, it was the equivalent of something like $484 billion dollars in today's values. But although wrong about that, on much else Barringer was uncannily right.

An abiding mystery of Meteor Crater has been the near absence of melted rock at the impact site. As we have seen, when meteors hit the atmosphere at cosmic speeds, the air is as solid to them as water seems to a belly-flopping diver. Huge stresses tend to break up any projectile, and although stony meteorites fragment more easily, even tough dense nickel-iron meteorites may be shattered. Planetary scientists Jay Melosh of the University of Arizona and Gareth Collins of Imperial College, London, have estimated that for the plummeting parent body of Cañon Diablo, its "crushing strength" would have been exceeded by the stresses acting upon it as it reached a height of about 9 miles above ground. From that point on, its fragments would have tended to spread out into an expanded cloud, about 219 yards across—more or less as Barringer came to believe. As drag increased on the fragments, one piece, containing roughly half the mass of its parent body, would have continued intact to the ground, colliding with it at about 7 miles per second—3 miles per second slower than its original cosmic velocity.

At impact, the collision would have released the energy equivalent of 2.5 megatons of TNT, while the equivalent of a further 6.5 megatons was imparted to the atmosphere. Because nickel-iron will melt at impact speeds of 6 miles per second, with complete melting at about 7 miles per second, it is likely that all of the main body was dispersed as droplets and vapor. This would explain the oxidized nickel that Barringer detected in crater sediments, and which were identified as frozen droplets of molten nickel-iron by the doyen of U.S. meteorite hunters, Harvey Nininger, in the 1950s.

However, the temperatures produced would have been too low to melt very much target rock, so its absence is not the problem it was once thought to be. Moreover, fragmentation of the parent body high in the atmosphere explains the wide strewn field of unmelted iron meteorite fragments around the site.

By the time Daniel Barringer died in 1929, $600,000 had been spent on prospecting operations at the crater—including most of Barringer's personal fortune. However, he did survive long enough to witness the triumph of the impact hypothesis, at least beyond the confines of the USGS. Barringer's son, D. M. Barringer Jr., continued to identify impact structures on the Earth, so that by the 1930s very few Earth scientists seriously entertained any other explanation of Barringer Crater. Gilbert himself never admitted his error, and his faithful survey remained fiercely opposed to impact cratering research until well into the 1950s. Yet we know that privately he entertained doubts. When Merrill was writing up his fieldwork for the 1908 paper, he wrote to the great man (in the autumn of 1907) to tell him about the new evidence he had found. Gilbert wrote in response: "The evidence that Tilghman and you have gathered inclined me strongly to the meteorite hypothesis . . . [but] What became of the meteorite?" Sadly, Merrill could do no more than hint that Gilbert's position had softened; but adherents of his 1896 view continued to dismiss the meteorite hypothesis. The damage was done.

The crater's impact origin was finally clinched in 1963, by USGS geologist Eugene Shoemaker. In 1961, Shoemaker had become founding director of the pioneering Astrogeology Research Program at the USGS in Flagstaff, Arizona, and had studied the craters left after nuclear tests at Yucca Flats, Nevada. There he and his USGS collaborator Edward Ching-Te Chao found a diagnostic variety of the mineral quartz, coesite, whose distorted crystal structure can only form under intense shock at high temperature. In 1960, Chao had been the first to identify this mineral in nature, and had named it after the high-pressure chemist Loring Coes who first synthesized it under laboratory conditions and published the results in *Science* in 1953.

Finding it at Meteor Crater, together with other characteristic shock features identified by the emerging field of impact
metamorphism, finally put the crater's origin beyond question
and vindicated the lifelong struggle of Daniel Barringer. The
spectacular discoveries of Chao and Shoemaker also redeemed
the USGS's reputation in impact geology. Its previous long
silence—which William Hoyt, author of the definitive book
on the controversy, described as an attempt "to preserve the
fictional infallibility that bureaucracies everywhere and at all
times have tacitly claimed"—had at last been broken.

Barringer knew the day would come and toward the end
had remarked darkly to his friends that whenever "his" meteorite theory was finally accepted, somebody else would probably get the credit. Sure enough, in 1928, *National Geographic*
published an article attributing the impact hypothesis to
Grove Gilbert. Barringer was incensed, but the magazine
wasn't wrong; Gilbert was indeed the first scientist to conceive
the idea, even though his mistaken conclusion and subsequent silence did more to hold back progress in understanding impact phenomena than any other single factor. The article
made no mention of Barringer at all.

It would be utterly wrong, however, to paint Gilbert as
some kind of villain in all of this. The British biologist and
science commentator Sir Peter Medawar wrote, in his book
Advice to a Young Scientist: "The intensity of a conviction that
a hypothesis is true has no bearing over whether it is true."
But, especially when pursuing an unpopular or revolutionary idea, a measure of bloody-minded conviction is necessary.
This battle between cool rationalism and conviction, between
reasoned and unreasoned belief—"faith" if you like, though
a scientist would call it "gut feeling"—is something that the

history of science yields again and again. Gilbert lacked that passion and was led astray by reason. On the other hand passion was something with which Barringer was, if anything, rather over-endowed.

It is significant, perhaps, that the Geological Society of America now makes a G. K. Gilbert Award each year to reward geologists judged to have made the greatest contribution to planetary science. No matter that he reasoned himself into error and found it impossible to back out of the cul-de-sac into which he had got himself, Grove Gilbert's first and best idea, so doggedly taken up and pursued by Barringer, remains as profound a leap forward in geological thinking as continental drift and plate tectonics.

Some have even said more profound. The story of Barringer Meteor Crater tells of the final recognition by mainstream Earth science that our planet lives within, and occasionally falls prey to, events in its cosmic environment—events that, as Derek Ager taught us, could achieve more in hours than all Earth's endless ages without name.

5

DEEP IMPACT

Many scientists unconsciously deplore the
resolution of mysteries they have grown up with and
have therefore come to love.

SIR PETER MEDAWAR, "LUCKY JIM"

For the last thirty years, scientists and public alike have largely accepted the view that the end-Cretaceous mass extinction, which did away with the dinosaurs, was caused by a giant impact. Many also believe that this extinction therefore had a single cause. And from these two ideas comes the assumption that all meteorite impacts must necessarily be bad news for life on Earth. All these ideas, which have become the new orthodoxy, are now being challenged in their turn. But although the revolution that established this orthodoxy seems with hindsight to have been almost instantaneous, it was not.

On April 2, 1985, for no apparent topical reason, the *New York Times* published an editorial on the end-Cretaceous mass extinction. It is hard to imagine any British newspaper doing this then, now, or indeed at any time; and even for the *Times* it must surely rank as an extremely rare event. Headlined "Miscasting the Dinosaur's Horoscope," the anonymous leader described how paleontologists, who, it said, had been struggling to understand the mass extinction for decades (not really true—most had been trying to ignore it), had had their ball snatched away by "two brash Berkeley scientists and a crowd of astronomers." The disciplinary trespassers in question were Luis and Walter Alvarez, a physicist–geologist father–son team who, together with Frank Asaro and Helen Michel, had introduced cat to pigeons in a paper published five years earlier. This paper suggested that the end-Cretaceous extinctions had been caused by a massive extraterrestrial impact.

The writer dismissed the very idea of meteorite impacts as a factor in life history, concluding that "the most immediate possible causes of mass extinctions" were to be found not in the heavens, but on the Earth. In a rhetorical flourish that doubtless roused cheers in conservative quarters, the editorial concluded by urging astronomers to mind their own business, and "leave to *astrologers* the task of seeing earthly events in the stars." Equating astronomers with astrologers is about as grave an insult as it is possible to imagine.

This attitude was looking distinctly outmoded by 1985. The piece accurately depicted the immediate knee-jerk reaction that many Earth scientists had had, about a decade earlier, when impact cratering first became widely discussed as a properly "uniformitarian" Earth-process on the long timescale of Earth history. Since then, things had quieted down; but by 1985 there was a resurgence of opposition. The revolutionary suggestion of Alvarez et al. had spread awareness of neocatastrophist thinking among a group who perhaps had been able to ignore it for longer—namely, vertebrate paleontologists.

Because of the massive popular appeal of their subject, vertebrate paleontologists and dinosaur specialists especially receive media attention out of all proportion to their numbers. A few even come to view themselves as stars and sometimes become more than a little preening about it. Happily, this is not the rule. Most vertebrate paleontologists, like their much more numerous colleagues working on less glamorous fossil animals without backbones, are cautious scientists who dislike pat solutions and are constitutionally unlikely to welcome any proposal offering a single, simple explanation of a complex problem like a mass extinction—which they had always regarded as "sudden" only in a geological sense.

As though to prove that the media do not control their scientific instincts, many vertebrate paleontologists were slow to embrace the death-by-impact idea even though it made dinosaurs even sexier than they already were. But there were plenty of non-scientific reasons for their dislike of the Alvarez hypothesis. For a start, it put them in a bad light. They, the professionals, had apparently been barking up the wrong trees for a century and more. It made them feel inadequate and, worse still, stole their thunder. Moreover, it didn't help that those who were making this heretical claim were not above spreading the imputation that the whole lot of them were poor scientists and little better than stamp collectors.

The philatelic jibe always rankles. Ever since it was coined by Ernest Rutherford, it has become a favorite rhetorical trick of physicists whenever they mount invasions of other people's subjects. Like the siren of the Stuka dive-bomber, its purpose is to engender panic. And maybe the brutal truth was, as many then reflected, if anyone was entitled to judge who was a good scientist and who wasn't, that person was surely Nobel Prize–winner Luis Alvarez. In fact this was not necessarily so. Although Alvarez was undoubtedly qualified to judge other physicists, whatever his belief to the contrary, this did not mean that he understood enough about how other sciences worked to make the same evaluation of its practitioners. However, this apparent arrogance got him into a lot of bad odor and helped paleontologists to believe mistakenly that they had seen all this before.

Dino-guru Bob Bakker, for instance, the revolutionary hero-pioneer of the warm-blooded dinosaur theory, was initially incandescent about the impact theory. He wrote, in 1985: "The arrogance of these people is simply unbelievable. They know next to nothing about how real animals evolve, live and become extinct. But despite their ignorance [they] feel that all you have to do is crank up some fancy machine and you've revolutionized science." It is, in fact, part of the folklore of geology that invading physicists, having marshaled their battalions of well-oiled and incontrovertible pronouncements, soon find themselves facing an icy retreat once they discover that the assumptions on which those pronouncements depended are after all mistaken. This most famously happened before the discovery of radioactivity when physicists insisted that the Earth was losing heat too rapidly to be as old as geologists wanted it to be. It happened again over continental drift, which physicists said was impossible—until their own evidence proved it was actually happening, when all the physicists' objections over its mechanism mysteriously ceased. And, of course, Sir Isaac Newton himself said that meteorites were unthinkable in his perfect cosmos.

But what paleontologists failed to notice was that this was not the same old story. This time, it was they and not the physicists who were saying "it's impossible." This threw people off balance for reasons that they barely perceived themselves, and the effect was highly disorientating. The *Times* editorial showed this effect rather neatly when it asserted: "invoking regular squads of meteorites to dispose of the dinosaurs and other vanished species is only to exchange one mystery for another." This was highly unfair.

To be sure, exchanging one mystery for another was a charge fairly leveled at many of the old theories that had been dreamt up to explain the dinosaurs' demise: including a majority that were impeccably Earth-bound. There was, for example, the notion that global warming so heated up dinosaurs' testicles that they became sterile; or that the rise of flowering plants had introduced alkaloid poisons into their diet, which dinosaurian tongues could not taste, nor their livers digest; or that the advance of this less fibrous flora removed roughage from their diet and killed the herbivorous dinosaurs by chronic constipation—leaving carnivorous *T. rex* with nothing to eat.

These theories fell firstly into the trap of only addressing the demise of dinosaurs—who had merely been the highest-profile victims of the end-Cretaceous mass extinction. This is like proposing a special theory for why so many movie stars are killed during an event that also wipes out 80 percent of the population of California. Yet their worst fault, and of many theories like them, was that they were fruitless; and they were fruitless because, while they were intriguing, even occasionally plausible, they were untestable.

It is always unwise to say "never," but it's a fair bet that we will never know what flavors dinosaurs' tongues could taste. Fossilized dinosaurian organs like livers and bowels have turned up recently in spectacular cases of natural mummification. But working out whether those silicified livers could metabolize alkaloids, or those stony bowels pass magnolia flowers less well than cycad leaves, will probably remain elusive. A good scientific theory, by contrast, generates its own tests. Dinosaurs may or may not have been castrated by global warming (probably not); but extraterrestrial impact at the end of the Cretaceous was a fertile, testable idea that arose from actual evidence. This simple difference made the impact hypothesis much more than just another mystery.

Like nearly all ideas, extinction by impact had been around for a long time, but had never been taken seriously. As long ago as 1694, Edmond Halley suggested to the Royal Society that the Biblical deluge may have been the result of "the casual shock of a comet," pushing the planet and sloshing the oceans about. Even closer to the mark, in 1742 French savant Pierre-Louis de Maupertuis suggested in a popular work called *Lettre sur le comète* that "comets have occasionally struck the Earth, causing extinction by altering the atmosphere and oceans." Independently, Pierre-Simon Laplace, whose theory of the origin of the Sun and planets within a solar nebula still forms the basis of modern scientific thought, wrote that "a meteorite of great size striking the Earth would produce a cataclysm that would wipe out entire species and destroy . . . all the monuments of human history."

Yet the combination of being (then) both untestable and unfashionable had meant that anyone suggesting extraterrestrial explanations could easily be branded a lunatic. Paleontologist Digby McLaren had made exactly that suggestion in a famous presidential address to the Paleontological Society of America, although about a different mass extinction—the biggest one of all, which occurred at the end of the Permian Period 370 million years ago. McLaren's 1970 address, entitled "Time, Life and Boundaries," was a staple of student reading lists, and I remember following orders to look it up sometime in 1975. Although intensive research over the intervening decades has failed to convince anyone that meteorites have had anything to do with any mass extinction other than that which killed the dinosaurs, McLaren's address's chief claim to posthumous fame now lies in those once much-derided final visionary paragraphs in which he seriously mooted the possibility of mass extinction by asteroid.

McLaren was not alone. Other extraterrestrial theories of extinction were also floating around at this time. The possibility that nearby supernova explosions might have bathed the Earth in radiation, so increasing genetic mutations that might have led to extinctions, was first suggested in 1963 by the highly respected German paleontologist Otto Schindewolf, in a paper whose title simply asked *"Neokatastrophismus?"* And the collapse of the Earth's Van Allen Belts due to reversals of the Earth's magnetic field also briefly became a fashionable extinction mechanism.

The Van Allen Belts, two doughnut-shaped envelopes containing charged particles, surrounding the Earth and held in place by the magnetic field, had been discovered by University of Iowa space scientist James Van Allen in 1958 after the early satellite missions Explorer 1 and Explorer 2. Scientific papers implicating them in extinctions first appeared in 1966, when reversals of the Earth's magnetic field were enjoying great notoriety. Although field reversals had been known about since the 1920s, in 1963 they turned out to be the key to proving that the ocean floor was being continuously created at mid-ocean ridges. As the ocean floor is formed and moves away from the ridge in opposite directions like two conveyor belts, the erupted ridge lavas freeze the prevailing magnetic field into their mineral structure. The discovery of symmetrical stripes of normal and reversed magnetic polarity in ocean-floor rocks either side of the mid-Atlantic Ridge by British geophysicists

Drummond Matthews and Fred Vine proved to doubting geo-physicists, with evidence from geophysics itself, that ocean floors spread, and hence continents drift.

Van Allen Belts had even entered the public consciousness, particularly though the sci-fi film and later TV series *Voyage to the Bottom of the Sea* (1961), which featured the nuclear submarine *Sea View*, commanded in the film by Walter Pidgeon, saving the world from being roasted to death by Van Allen Belts "catching fire." Everyone loves a new idea in science, especially if it makes an early leap to celluloid. It would not be long before this same cultural phenomenon would get behind the end-Cretaceous impact bandwagon. However, the collapsing Belt theory itself collapsed when it was realized that any consequent increase in radiation would hardly have been enough to ripen a tomato.

Three years after Digby McLaren tried to explain the end-Permian extinction by meteorite, in 1973 the great geo-chemist Harold Urey made the same suggestion about the end-Cretaceous extinction itself. He had found a layer of tektites at the base of the Cretaceous. Tektites are blobs of solidified natural glass, formed when molten rock, sprayed up after an impact, flies through the atmosphere, congeals in flight and strews large areas with enigmatic greenish beads with weird aerodynamic shapes. In a paper called *Cometary Collisions and Geological Periods,* Urey hit the jackpot straight away, correctly suggesting that an impact had occurred at the end of the Cretaceous.

Perhaps it was an idea before its time, but no worldwide effort to look for more tektites at the end of the Cretaceous was initiated by Urey's paper. Somehow, the world of science in 1973 found it possible to ignore his simple glass objects, yet in 1980 it found it impossible to ignore new evidence of the

same event. Timing was one factor, as was the more technical nature of the new evidence. And so was the shock-and-
awe effect of having a Nobel-prizefighter like Luis Alvarez on
the team, with his knack for PR and famous disregard for the
finer feelings of others, especially those in his way. And any
anti-physics prejudice was, for most geologists, considerably
weakened by the geological credibility that was lent by the dis-
arming presence of Luis's distinguished geologist son Walter,
with his ready charm and evangelizing smile.

✳

The 1980 Alvarez paper neatly tied together evidence from two
different aspects of geological science—the study of microfossils, which allows rocks to be accurately dated and correlated,
and geochemistry, whose analyses indicated the sudden arrival
of an extraterrestrial object. Moreover the famous roadside
section where Luis and Walter Alvarez first made the connection and found the evidence for an end-Cretaceous meteorite
impact lies a mere 62 miles east of Siena, in the Bottaccione
Gorge, near Gubbio.

We have already seen how, in June 1794, stones rained on
Siena, and how for the first time it was the Academy, rather
than the peasantry, which witnessed a meteorite fall. We also
met Abbé Ambrogio Soldani, perpetual secretary of Siena's
Accademia dei Fisiocritici, who collected witness testimony
and sent samples of the meteorite to the exile Guglielmo
Thomson. Soldani went as far as believing that the stones had
indeed fallen, just as peasants had been saying for centuries,
and hypothesized that they were the product of earthy and
metallic dust, coming together in a high cloud. He stopped
well short, however, of sourcing the stones in space. Soldani's

other claim to scientific fame was as a founder of micropaleontology—the study of microscopic fossils—which was to prove
so vital to Alvarez et al. in identifying and then dating the
traces of an end-Cretaceous meteorite impact.

The Accademia dei Fisiocritici, still in its original premises on the Piazzetta Silvio Gigli, is perhaps the greatest
unsung glory of Siena. Although you will struggle to find it
in guidebooks, this perfectly preserved scientific institute,
which dates from the last years of the seventeenth century,
is open to the public. You will have to ring the doorbell, but
do so. Inside, around an airy courtyard, you will be able to
walk Soldani's cloisters and examine the beautiful models
and specimens of rocks, minerals, animals, plants, and fungi
amassed by its keepers over 400 years. The cabinets, with
their handwritten labels, do not show the slightest sign of
neglect. Although it is—as Dylan Thomas once wrote—"a
museum that ought to be in a museum," for the Fisiocritici
the description is no insult.

In a side room just off the ground-floor cloister stands a
marquetry case containing the smoothly sliding wooden
drawers that since 1871 have held Soldani's bequeathed collection of fossil foraminifera—the tiny, intricately chambered
shells made by single-celled organisms, collected from the
rocks around Siena. In the same vitrine, on a black wooden
plinth, stands a piece of the *Meteorite di Siena*—or at least, a
piece of one fragment from the fall that descended on nearby
Lucignano d'Asso and one of the few that remain after Soldani
sent samples of it all over the world.

Fossil foraminifera are incredibly useful tools for telling
the relative ages of the rocks in which they occur. Moreover,
although they survived to the present day, the foraminifera

took the events of 65 million years ago very badly; the end of the Cretaceous being officially marked by (among other things) the disappearance of one of their number called *Globotruncana*.

✳

Geologist Walter Alvarez, who with his father in 1980 made the rock sections of Bottaccione Gorge famous, first encountered the gorge on the day after Christmas a decade before, during a blizzard. Earlier that day, snow and ice had lent charm to Assisi and the basilica of San Francesco; but not to the treacherous cobbles and firmly closed doors of medieval Gubbio. Alvarez recalls the legend of how St. Francis himself had once gone to Gubbio to entreat a wolf to stop eating children. It seemed to him and his wife, Milly, that contrary to legend the saint had failed. They left Gubbio without regret and, following a hopelessly out-of-date map, decided to make for Ravenna via the Bottaccione Gorge. They found the pass blocked and were forced back. It was an inauspicious start.

The rocks exposed in the walls of the Gorge span the end of the Cretaceous and the beginning of the Palaeocene periods. What brought Alvarez back in 1973 was the pink color of those limestones because pink usually means one thing: plenty of iron; and iron is magnetic. Alvarez's research at that time was aimed at shedding light on the perplexing geological history of the Apennine mountains, and the strange and complex tectonic processes that are building them as Africa moves north against Europe. By carefully drilling oriented cores from these rocks, Alvarez and palaeomagnetics expert Bill Lowrie hoped to take samples that would show how the various geological terrains composing the Italian peninsula had rotated as they were folded and uplifted.

Although that project was not so successful, in the process Alvarez and Lowrie discovered that the limestones—called Scaglia Rossa—recorded some of the Earth's periodic magnetic reversals. He and Lowrie had first met as young researchers at the Lamont-Doherty Geological Observatory at Columbia University, which was then at the epicenter of the Earth-shaking discoveries of the plate tectonic revolution. They knew that magnetic reversals were now a hot topic, and that a major problem facing geologists at the time was how to extend the record of these reversals backward in time, into rocks older than a mere few million years.

Magnetic reversals had provided geophysical evidence that continental drift was real, and that oceans like the Atlantic were widening. The age of each magnetic reversal recorded in the volcanic sea-floor rocks could be determined by dating the lowermost sediment lying on top of them; but for older rocks, this was rather difficult, and for much older rocks impossible because all ocean floor over a certain age has been destroyed at subduction zones, where it plunges back into the planet again. Alvarez and Lowrie's discovery in the Bottaccione Gorge offered the hope of dating some ancient reversals much more easily and cheaply than drilling through thick ocean floor sediments to basaltic rocks beneath, and of relating those reversals to other events in geological time.

The limestone sequence laid out along the roadside could be dated very precisely by its contained microfossils (Soldani's foraminifera), so any magnetic reversals they found could be fitted neatly into the existing framework of Earth history. The Scaglia Rossa is almost 437 yards thick, and the first task before taking oriented drill-samples was to measure the section precisely. The rocks had been tilted to 45 degrees since they had been laid down, so walking up the gorge conveniently took the geologists into progressively younger rocks, and eventually to the top of the Cretaceous, at 380 yards in their measured section. There, where the foraminifer *Globotruncana* vanished, was the K-T Boundary, visible as a thin band of clay barely an inch thick. Alvarez and another co-worker, Terry Engelder, took the crucial sample of that clay in 1977.

Clay partings between the limestone beds are common in the Scaglia Rossa. This parting was thicker, but the only other unusual thing about it was its location. This was not just any division between geological periods. The Cretaceous was the last period in a much longer span called the Mesozoic—the era of "middle life" when, as they say, dinosaurs ruled the Earth. Above lay the Cenozoic—the era of "recent life," in which we live now—the era of mammals. This was therefore a specimen of almost mystical significance. How long had this layer of clay taken to be laid down? On that, all depended. If its deposition had been slow, the extinction event would also have been slow, and its suddenness only apparent—a product of reduced sedimentation rate. This was a hypothesis of which Charles Lyell would have approved. But if the reverse were true, and sedimentation of the clay had carried on at normal rates, then that would mean the extinction event had been genuinely fast.

But how fast? Geologically fast (less than a million years, say), archaeologically fast (less than a thousand years) or human-life fast—last Tuesday afternoon? To narrow this down, Walter needed to find an accurate measure of depositional rate, and that quest brought him back into contact with his illustrious father, Luis. Walter says that he had not hitherto had very much contact with his father. After leaving home for college, Walter had lived the peripatetic life of a geologist—in Holland,

Libya, and Italy—and hadn't really had the chance. However, when Walter got a job at Lamont-Doherty in 1971, his return stateside provided the means. Walter's scientific need provided a ready and urgent motive for a reunion, and the opportunity finally came when Luis flew over from Berkeley on a visit. For the first time, Alvarez the elder became interested in geology and the amazing excitement being generated by the plate tectonic revelations that were just reaching their peak.

Six years later, Alvarez the younger moved again—this time to his ultimate academic resting place, Berkeley—the very university where his father also taught. "It was a stroke of such good fortune that even now I can scarcely believe it happened," Walter wrote later. Luis was recently returned from Egypt, where he had been using cosmic radiation to X-ray the Pyramids (as you do). The method had worked, but disappointingly had revealed no lost treasure chambers—so Luis was eager for a new project. How long had the K-T Boundary clay taken to deposit?

Luis Alvarez quickly conceived a way to answer this question. His first idea was to measure the concentration of the radioactive element beryllium-10, which is produced continuously in the atmosphere and falls to Earth at a constant rate. The more of it they found in the clay, the longer it had taken to deposit. Sadly this method failed because amazingly the published half-life of beryllium-10 was wrong—the unstable isotope decayed too quickly to be useful. Luis Alvarez knew he needed another timepiece, something that accumulated always and everywhere at the same rate. He hit on the element iridium.

Iridium is element number 77, a rare metal and the second densest element known after its neighbor osmium. White

like platinum but tinged with yellow, iridium is so resistant
to corrosion and high temperature that it tends to be used in
such things as fountain pen nibs and spark-plug electrodes as
well as certain light-emitting diodes in flat panel screens. The
world's mines produce only about three tons of iridium every
year, and society eagerly uses all of it in these specialized but
crucial applications.

Everywhere in the Solar System, chemical elements occur
in proportions that became more or less fixed even before the
planets formed, and for this reason, Earth as a whole contains
the same amount of iridium as any other product of the solar
nebula—which is to say, about half of one part per million. But
that is an average whole-Earth figure. Because the Earth is geologically active, different elements have been segregated over
time into different areas—some concentrating in the crust, others in the mantle and core. The crust is much poorer in iridium
than the planet as a whole, containing only about 0.001 parts per
million. The difference is due to the fact that the Earth's internal
heat caused most of its nickel and iron to melt and separate out
to form the core, long ago. When that happened, many elements
like iridium, which are siderophile (iron-loving), were more soluble in the molten metal than silicate rock, and sank down into
the planet's center.

Most meteorites, by contrast, come from parent bodies that
did not undergo this kind of differentiation, and still maintain
nebular abundances of all elements. All in all, there is a lot of
iridium floating around in space—certainly a lot more than
there is on the surface of the Earth. If the K-T clay recorded
a slowdown in the rate of deposition, it would tend to show
slightly elevated concentrations of iridium. Conversely, if
overall sedimentation rate had been higher, one would expect

to find a lower than normal concentration—possibly even beyond the limit of detectability.

What Frank Asaro found when he did the analysis on the boundary clay layer from Gubbio were concentrations that were neither lower nor slightly raised. What he found astonished him and would soon astonish the Alvarezes and the world. Asaro initially reported finding three parts iridium per billion. Against average crustal background values, this was not just high—it was enormous. Quickly, the experiments were redone to test for experimental error. They found one error; but its effect had been to reduce, not increase, the result. The final concentration Asaro determined was nine parts per billion—90 times higher than the Alvarezes' highest expectation. They had discovered the now famous iridium anomaly.

The obvious next step was to look at rocks of the same age somewhere else in the world, to see if this iridium anomaly occurred there too. The site they chose was a cliff section south of Copenhagen, Denmark, called Stevns Klint. Once again, they found a clay layer, and once again there was a peak in iridium values.

Could Otto Schindewolf have been right? people asked briefly. Could his idea of a nearby supernova event irradiating the Earth and killing off the dinosaurs have been correct all along? Supernovas, as we have seen, are prodigious generators of the heavier elements in the periodic table and could perhaps have showered the Earth with iridium. But if that had been the case, there would be other elements too, and Luis Alvarez, the nuclear physicist, knew that one of these would have been plutonium-244. To test for it, they brought onto the team the plutonium chemist Helen Michel (pronounced "Michael"), who found no plutonium. Schindewolf's supernova hypothesis was finally exploded.

Conclusively disproving a hitherto untestable idea was not a poor outcome for the Alvarez team; but in science, positive results are the real deal. Over the summer of 1979, while Walter was back in Italy continuing his palaeomagnetic researches, Luis began to develop ideas about how a gigantic extraterrestrial impact might have created a worldwide iridium anomaly, killing ammonites in the ocean as effectively as the dinosaurs on the floodplain and all the other things that perished at the end of the Cretaceous. What he came up with was an idea that not only changed our view of Earth history but arguably helped change history itself: dust, darkness, and death.

As luck would have it, a major international meeting on the K-T extinction was being held in Copenhagen that September. Luis phoned Walter from Berkeley and urged him to present the iridium data there, ruling out the supernova and ruling in a meteorite impact. Walter remembers how sure his father was that everyone at the meeting would be absolutely delighted to have their mystery solved for them.

Of course, as Walter suspected, it didn't quite happen that way, and paleontologists as a group were decidedly ungrateful at first that their beloved mystery should have been apparently solved for them by strangers. Yet the meeting was significant for another reason—it brought Walter together with a young Dutch researcher named Jan Smit.

The Alvarezes' impact idea had already hit the popular press, as both father and son had spoken to the media. At that time, Smit was finishing his Ph.D research on rocks near Caravaca de la Cruz in southern Spain, which also cover the K-T Boundary. He read about Gubbio in the British science magazine *New Scientist* and immediately went to check his data from his own geological section. What he found in his element-abundance data confirmed the findings of Alvarez, Alvarez, Asaro, and Michel.

Smit's work had, in fact, pre-dated the Berkeley team's; though because he had been ill with glandular fever, he had simply lacked the energy to look them over—until the *New Scientist* article drove him to it. Ever since, Walter Alvarez has regarded Jan Smit as co-discoverer of the K-T iridium anomaly, and hence godfather to the idea that dinosaurs succumbed, at least in part, to death from above—during what we now know was the only mass extinction event in Earth history to have coincided with a meteorite impact.

But nobody knew that then, and hopes rode high among many scientists that perhaps a single cause had been found not just for one, but for all mass extinctions—one that chimed well with the nuclear terrors of a newly nervous age. Asteroids and comets, stalking the darkness beyond the Earth, became geologists' boogeymen of choice.

6

DESTROYERS OF WORLDS

Most hot ideas turn out to be wrong.

STEPHEN JAY GOULD

Bloomsbury, that elegant area of London surrounding the British Museum and the University of London, is a very intellectual bit of real estate, keen to make visitors aware of its place in cultural history. Yet there is one corner of Russell Square which to this day remains completely unmarked—despite that it arguably outweighs in significance all the others, marked and unmarked, many times over. Near the Imperial Hotel, on Southampton Row, stands a set of traffic lights. At that spot, on September 12, 1933, the possibility of a nuclear chain reaction was first conceived in the mind of man.

The man in question was dapper émigré physicist Leó Szilárd. Blessed with sound political intuition, that same year Szilárd had seen Hitler appointed chancellor of Germany on January 30, the Reichstag set ablaze on February 27, and the dismissal of Jewish academics from German universities on April 7, but unlike so many others he knew what to do and had the means to do it. By May, he was in London, living on his modest personal capital and working to find employment for his displaced colleagues all over the rapidly shrinking free world. For many, Szilárd included, this trauma would be the making of them; they would flower more spectacularly in their new habitats than they would ever have done at home. But such success lay, at that time, far in the future.

Szilárd hadn't much time to think about physics at this frantic juncture in his life, but reports of the meeting of the British Association for the Advancement of Science, taking place that year in Leicester, reached him through the pages of *The Times* as he sat in the lobby of the Imperial Hotel. Today, the Imperial Hotel is a modern concrete-and-glass structure quite different from the *fin de siècle* fantasy of old, which more closely resembled the still extant Russell Hotel nearby. In the comfortable gloom of its lobby, Szilárd read a series of stacked headlines above a dense column, as was the newspaper fashion then:

THE BRITISH ASSOCIATION

*BREAKING DOWN
THE ATOM*

*TRANSFORMATION OF
THE ELEMENTS*

There followed an account of a lecture by Ernest, Lord Rutherford, the first man to split the atom. Rutherford told the folk of Leicester that in the future, while it may be possible to transform one element into another by bombarding its atoms with protons (positively charged subatomic particles), the process would not yield any energy. In all probability, it would cost energy. Looking for power from the atom, he said, was "moonshine."

Szilárd had a rebellious dislike for dogmatic statements and, feeling a little disgusted, decided to go for one of his long walks. He admired and had already got to know H. G. Wells, author of *The Shape of Things to Come,* then newly published, and whose 1914 novel *The World Set Free* had foreseen a world of atomic power and warfare. He wrote later: "I remember that I stopped for a red light at the intersection of Southampton Row . . . I was pondering whether Lord Rutherford might not prove to be wrong."

Szilárd knew that neutrons, with the mass of a proton but no charge, might be persuaded to hit an atomic nucleus quite easily. This idea was not new; but then Szilárd remembered about chemical reactions in which one interaction between molecules generated two more, each of which gave two more, and so on in geometric progression until all the reagents were used up.

"As the light changed to green and I crossed the street it . . . suddenly occurred to me that if we could find an element which is split by neutrons and which would emit two neutrons in the process . . . such an element, if assembled in sufficiently large mass, could sustain a nuclear chain reaction." This would mean that the sufficiently large mass could, if unmoderated, be transformed into its equivalent in energy. By the time Szilárd stepped on to the opposite sidewalk, the nuclear age had begun.

Physics soon found itself transformed from an arcane, low-budget pursuit, joining chemistry as part of the military-industrial complex that provided weapons for governments and received massive research-funding in return. The First World War had been chemistry's conflict, fueled by the Haber Process for the artificial fixation of nitrogen, essential ingredient of conventional high explosives. The Second World War would be physics' big break. The realization that the human race could now commit collective suicide—a point when our species could be said to have come of age—began on that narrow crossing in Bloomsbury. It was a Tuesday.

The term "atomic fission" did not come into use until 1938, when Otto Frisch and his aunt Lise Meitner became the first scientists to observe the uranium atom splitting under neutron bombardment, accompanied by the release of stupendous energy in the form of neutrons. Szilárd's exponential chain reaction became a reality. The distinguished scientist Rutherford had indeed been wrong, and science fiction writer H. G. Wells by contrast had been right. Atomic energy and nuclear war were now reality.

Crucial experiments confirmed the facts in January 1939. Science journalist Thomas Henry wrote them up for the *Washington Evening Star's* Saturday afternoon edition on January 28. Over the wires, Henry's story reached the *San Francisco Chronicle*, whose Monday edition fell into Luis Alvarez's hands while he was having his hair cut in Stevens Union, Berkeley. Haircut half finished, Alvarez dashed back to his laboratory. After wiring colleagues for more information, Alvarez the elder took the news to another Berkeley physicist, J. Robert Oppenheimer.

Oppenheimer was the charismatic genius who eventually oversaw the Manhattan Project, on which Alvarez also worked, and which led to the building of the two first (and perhaps last) atomic bombs to be dropped in war. One of Oppenheimer's students, Philip Morrison, later recalled that "within perhaps a week" of hearing Alvarez's news, Oppenheimer's blackboard contained a very bad drawing of a bomb.

The two fission bombs, Little Boy and Fat Man, used different elements to create their explosions. Little Boy used uranium; Fat Man plutonium. The uranium weapon was easy to set off—all you had to do was fire a slug of uranium into a larger mass of the same metal, which then became sufficient, or critical as we now say, to produce Szilárd's exponential chain reaction. Little Boy was dropped on Hiroshima.

However, the same simple detonation method was not available for plutonium, which had instead to be explosively compressed—an enormous technical problem, part of which was getting the implosion to occur evenly over the whole surface of the bomb assembly. Focusing the pressure with lenses of shaped explosive was only part of the problem. The other was setting off the many separate charges at precisely the same time. After working on this problem for two years, Luis Alvarez solved it with an exploding bridgewire detonator mechanism that could fire caps simultaneously to an accuracy of millionths of a second. Alvarez also invented a set of parachuted sensors that could measure the strength of the explosion and help calculate the bombs' yield. An aviator himself, Luis eventually flew with the B-59 *The Great Artiste,* which tailed the *Enola Gay* to Hiroshima on August 5, 1945, and watched the realization of Oppenheimer's nightmare, first glimpsed at the test site Trinity in the deserts of New Mexico, when the line from the ancient Sanskrit epic the *Baghavad Gita* had run, according to a documentary 20 years later, through Oppenheimer's mind:

All at once I am become death, destroyer of worlds.

After the success of the fission weapons and the swift end they brought to the Pacific theater of the war, America looked forward to a post-war world in which it enjoyed a nuclear monopoly. But it was not to be. In September 1949, news reached Oppenheimer that U.S. planes flying over Japan had detected tell-tale radioactive particles that could only mean one thing—the Soviet Union had tested a fission weapon. Pressure grew for the development of an even more powerful fusion weapon—the hydrogen, or thermonuclear bomb—which brought to Earth the power of the stars themselves.

"The super," as it was then called, required a successful fission explosion to set off the fusion of hydrogen atoms to make helium, releasing energy according to Einstein's famous ratio $E=mc^2$—or, energy equals mass times the speed of light multiplied by itself. This project had been a glint in the eye of Edward Teller since 1942, when the Manhattan Project was still developing fission bombs. Oppenheimer, as a project manager anxious not to divert effort, did what he could to keep the super on the back burner. After the war, he continued to oppose the project—believing that if America refused to ratchet the arms race another notch, the Soviet Union wouldn't either. Teller, who is commonly assumed to have been the model for Stanley Kubrick's *Dr. Strangelove,* thought this naïve, and Alvarez supported him. Despite one false move, where Luis was apparently bamboozled by Oppenheimer's political subtlety into signing a report delaying the H-bomb, Alvarez never wavered in his belief that while nuclear weapons were regrettable, the arms race was inevitable.

A historical irony emerges in the tale of Luis Alvarez's involvement with the development of nuclear weapons—for by discovering the iridium anomaly and conceiving the idea that a single meteorite strike might have caused mass extinction, it soon led others to realize that a global thermonuclear war would have similar effects—and might not be survivable at all. This notion of nuclear winter, as it became known, became the strongest weapon in the campaign to eradicate the very weapons that Luis Alvarez helped develop.

The first step on that journey took place two years after the publication of the original paper on the iridium anomaly, in 1982. In that year Luis Alvarez gave a speech in which he calculated the energy released by an impacting 6-mile-diameter asteroid and compared it with an all-out nuclear war—clearly implying that such an interchange would have similar consequences. The Cold Warrior's language of megatonnage began to enter the impact scenario; but the compliment was returned. The theorists of nuclear conflict had thought exclusively about the immediate consequences of their weapons—casualties, megadeaths, social cohesion, fallout. The main assumption had been that after not so very long, things would return to normal. No one had thought to ask what the long-term environmental effects of a global nuclear war would be. The answer came that very year in a paper by Dutch atmospheric chemist Paul Crutzen of the Max Planck Institute for Chemistry in Mainz, Germany, and John Birks—a chemist specializing in the kinetics of chemical reactions at the University of Colorado. The title said it all: *The Atmosphere after a Nuclear War: Twilight at Noon.*

Crutzen shared a Nobel Prize in 1995 for work on the depletion of the atmosphere's ozone layer, and told in his Nobel lecture how in 1981 the editor of the journal *AMBIO* asked him to contribute to a special issue devoted to the environmental consequences of nuclear war. The resulting study was the first to apply computer simulations of the atmosphere to this problem and the picture it painted closely resembled the Alvarez end-Cretaceous extinction scenario, to which Crutzen and Birks made direct reference.

This research idea was taken up in 1983 by astrophysicist Carl Sagan, well-known secular humanist and science

The "Death Mask of Agamemnon," found by Heinrich Schliemann at Mycenae in Grave V, Grave Circle A in 1876. Later work has proved that it is 400 years too old to be what Schliemann thought it was.

Professor Derek Victor Ager (1926–1995).

Luis and Walter Alvarez, pictured at the K-T Boundary (just above Walter's hand), Bottaccione Gorge, near Gubbio, Italy.

Barringer Meteor Crater, Arizona, USA.

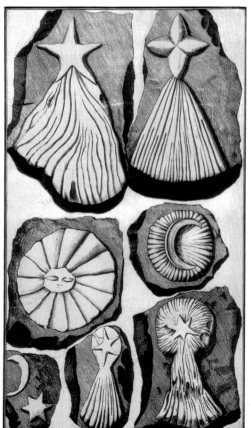

Plate from Beringer's Lithographiae Wirceburgensis, *illustrating some of the "figured stones," including four fine fossil "meteors," or "comets," complete with incandescent tails, which were maliciously made and planted for him to find.*

Jean-Baptiste Biot (1774–1862), Professor of Mathematical Physics at the Collège de France.

Mendicant physicist, musician, and inventor, Ernst Florens Friedrich Chladni (1756–1827), "father" of meteoritics (and acoustics).

Dangerous liaison. Antoine-Laurent de Lavoisier (1743–1794), pictured with his bride, Marie-Anne Pierette Paulze.

Varius Avitus Bassianus (ca.203–222), aka Roman Emperor Elagabalus, high priest of the meteorite.

Grove Karl Gilbert (1843–1918), chief geologist of the United States Geological Survey.

The Ensisheim meteorite, which fell near the Alsatian town on November 7, 1492.

The now demolished Imperial Hotel, Russell Square, Bloomsbury, London, from an old postcard. The corner on the right is where the idea of an atomic chain reaction first occurred to Leó Szilárd.

Above left: *A young boy who was struck by a meteorite in 1992 while playing soccer in Mbale, Uganda. Luckily only one of the smaller fragments—weighing about 3 grams— hit him, after bouncing through foliage.*

Above right: *The Irish-Canadian paleontologist Digby Johns McLaren (1919–2004).*

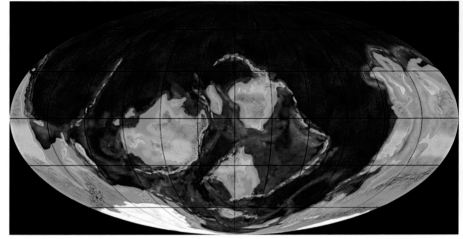

The mid-Ordovician world. A single mega-continent called Gondwanaland, comprising modern-day Africa, South America, Australia, Antarctica, and Arabia, stretched from the southern ocean to a few degrees north of the equator. Several smaller continental fragments include, nearest to the Pole, Baltica, comprising Scandinavia and much of northern Europe (west of the Ural Mountains, which had not yet formed). Straddling the equator is Laurentia (mainly modern-day Canada/North America, and Greenland). Further east, another continental island comprises modern Siberia.

Above left: Estonian astronomer and astrophysicist Ernst Julius Öpik (1893–1985).
Above right: Lembit Öpik, grandson of Ernst Julius, former Liberal Democrat member of parliament for Montgomeryshire, UK, who championed the Spaceguard project in the UK parliament.

La Nona Ora *(The Ninth Hour) by Italian artist Maurizio Cattelan.*

Abbé Ambrogio Soldani, Perpetual Secretary of the Accademia dei Fisiocritici in Siena.

Sheffield geologist and pioneering microscopist Henry Clifton Sorby (1826–1908).

Major Edward Topham (1751–1820).

Monument, erected on the spot where the Wold Cottage meteorite fell in 1795 by Edward Topham—to "perpetuate my credulity to posterity."

The Hoba meteorite, Namibia—estimated to weigh about 60 tons, it is the largest meteorite in the world and the largest naturally occurring iron mass on the surface of the Earth. It was discovered in 1920, buried by soil, and had to be dug out—hence the apparent "crater" in which it sits. It is thought to have fallen less than 80,000 years ago.

popularizer. Sagan was then perhaps the leading public man of science, as influential with the liberal left as Teller was with the upper echelons of the conservative right. He and four colleagues performed yet more computer modeling, their combined initials (Turco, Toon, Ackerman, Pollack, and Sagan) becoming the acronym TTAPS for the report they wrote that *Science* published.

The early 1980s were a time when the Cold War was being revived, thanks to the presence in office of two more than usually bellicose leaders: Ronald Reagan in the White House and Margaret Thatcher in Downing Street. Thatcher supported NATO's decision to put Pershing cruise missiles in Europe; deterrence, rather than détente, was their game; permission to site cruise missiles at the Greenham Common airbase led to mass protests as the British fleet was re-armed with new US-derived Trident missiles. Thus began the so-called "second Cold War," from 1979 to 1985. As the language worsened, popular anxiety grew.

Nuclear winter theory, for all that it was criticized as sloppy science, and for all that it cannot be linked directly to subsequent political changes, had the effect of rendering a nuclear war—even in limited theaters—unthinkable. Soviet president Mikhail Gorbachev, asked why he was warning of the dangers of nuclear war at this time, said: "Models . . . showed that a nuclear war could result in a nuclear winter that would be extremely destructive to all life on Earth. The knowledge was a great stimulus to us, to people of honor and morality, to act." In Britain, the government's pathetic civil defense advice, received in such public information campaigns as *Protect and Survive* (close windows and curtains, take kitchen door off hinges and stay under it for two days), was already almost universally ridiculed. It looked even more absurd when people realized that survivors of a nuclear war would face nothing but a new Dark Age. The lights of civilization would go out, perhaps forever, under the combined onslaught of cold, dark, disease, starvation, chaos, and fear. Raymond Briggs, the graphic novelist and creator of *The Snowman*, described it in his 1982 book *When the Wind Blows*. The same horrific scenario was written into influential live-action films, like Nicholas Meyer's 1983 depiction of the effects of war on Lawrence, Kansas, *The Day After*, as well as the BBC docu-drama *Threads* (1984), set in Sheffield. All carried the same message—those who die will be the lucky ones.

Paul Crutzen, accepting his Nobel Prize, said of his 1982 paper: "Although I do not count the "nuclear winter" idea among my greatest scientific achievements . . . I am convinced that from a political point of view, it is by far the most important because it magnifies and highlights the dangers of a nuclear war and convinces me that in the long run mankind can only escape such horrific consequences if nuclear weapons are totally abolished." And so Luis Alvarez, present at the conception of atomic and thermonuclear weapons, present at their terrible birth, supporter of their development through the first Cold War, had helped the world discover that the simultaneous release of such huge quantities of energy into the atmosphere would most likely send us the way of the dinosaurs.

After a new phenomenon is discovered and becomes established, it is rarely long before people start seeing it everywhere and looking for patterns. Perhaps this is all part of the phenomenon of having to believe in something before it becomes visible. In a very complex system like the Earth, it can be all too easy for the human mind, which wants nothing more than to see patterns, to be deceived. This is not to say that cycles do not exist; just that they require rigorous and impartial mathematical proof before gaining even one rung on the ladder to general acceptance. Back in 1973, one of Derek Ager's research students, Chris Walley, now an award-winning geology teacher in South Wales, pointed out to his supervisor a possible cycle in mass extinctions of 90 million years. You should always be careful what you say to your research supervisor because he might well publish it as a "personal communication" and credit you—which is indeed what happened to Walley.

However, the idea of cyclicity in mass extinctions took on feverish proportions ten years later in the 1980s, with publication of a rigorous mathematical analysis of a massive dataset of paleontological extinctions among marine vertebrates, invertebrates, and protozoans (single-cell organisms, like the foraminifera). The data had been laboriously gleaned from published information by the late Jack Sepkoski. Using computers, Sepkoski and Dave Raup, also of the University of Chicago, discovered a statistically significant cyclicity with an interval of 26 million years. Being unable to find any credible earthly explanation for such a periodicity, they speculated— because by then not only was it no longer heretical to do so but positively fashionable—that some extraterrestrial cause might be responsible. At that time scientists believed that two mass extinction events might be impact-linked. Perhaps all of them were. Could there be some mechanism whereby the Earth might find itself in the cosmic firing line at regular intervals?

It was not long before other scientists, one of whom was Walter Alvarez, claimed to have found a similar periodicity in the ages of known impact craters. Through the pages of the magazine *New Scientist*, in a piece written by my old friend and mentor, geologist-journalist Richard Fifield, the idea found its way to a general audience—including astrophysicists. Before long, they had developed a few ideas about how the Earth might be periodically pelted; the most popular of these involved a companion star to the Sun, an unidentified stellar object orbiting our own at about two light years' distance. This is not as impossible as it sounds, though if true it would be by far the closest star to our Sun. Some stars are dim and hard to see, and a surprising number of stars of all kinds are still unidentified. It all might sound a bit B-movie, but it was all possible. Or at least not impossible.

The companion star was quickly dubbed Nemesis—though there are mythological reasons why it would have been more appropriate if it had been named for the Hindu deity Shiva, since people who suffer the attentions of Nemesis have done something to deserve it. The orbit of this star could bring it, every 26 million years, close to where comets live.

The history of thought reveals that there are very few, if any, ideas that have not occurred already to someone in the past. Just like meteorite falls, ideas occur all the time, but only develop significant meaning as a result of who witnesses them and when. In science, the person who thought of most things first turns out usually to be either the Roman philosopher Lucretius—or even older and Chinese. In the matter of comets and their origins, pre-dating every scientist of modern times by a long way was an Estonian astronomer and astrophysicist by the name of Ernst Öpik.

Apart from his many scientific accomplishments, Öpik was patriarch of the greatest scientific dynasty ever to come out of the tiny Baltic state. He was more than just scientifically fertile, fathering three daughters by his wife Vera, as well as a son Uno and two more daughters by his research assistant Alide Piiri. One of his daughters, Helgi Öpik, is a distinguished plant physiologist (who, as it happens, taught me at university). Ernst's brother Armin Öpik was an eminent paleontologist, and Ernst's grandson Lembit Öpik was, until the 2010 election, a Liberal Democrat MP. Lembit, who enjoys some tabloid notoriety in the UK for his liaisons with glamorous weather girls and pop divas, will have his part to play in this story a little later.

Ernst Öpik's father had served in the Imperial Russian Navy, and he himself had volunteered for service in the White Russian Army. So the arrival of the Soviets in Estonia in 1940, followed by the Nazis and then the Red Army again in 1944, finally drove him to put down roots in pastures new. Fleeing first to Germany, he later settled in Northern Ireland, at the Armagh Observatory, where he worked for the remainder of his long career.

Öpik's astronomical interests ranged widely, but his main speciality was the minor bodies, the asteroids and comets. In 1932 he first conceived the idea that comets originated in an orbiting cloud at the farthest edge of the Solar System. This idea was arrived at independently eighteen years later by the Dutch astronomer Jan Oort after whom the notional cloud—which has still not actually been observed—is most usually named.

Öpik supposed that the loosely bound orbits of the cloud's cometary bodies could become unstable over long periods, and that periodically these icy objects were dislodged and sent on new trajectories that brought them into the inner Solar System. It also occurred to him that impacts from such bodies on the

Earth might well have happened in the past, and could have affected the history of life. Interestingly, though, Öpik was concerned less with impacts causing global devastation and mass extinction and more with those that had more regionally damaging effects.

If a comet impact devastated only a certain zone, which he called the lethal area, it could, for example, kill off an entire species if it were endemic only to that area. Thus, if the impact happened in Australia, for example, it could exterminate the kangaroo—but not all marsupials, which live in many other parts of the globe. Much of animal and plant distribution on Earth is provincial in this way, and for such organisms extinction by impact need not demand a global catastrophe.

While such limited theater extraterrestrial impacts may be insufficient on their own to cause a mass extinction like that seen at the K-T Boundary, the notion of sterilizing large areas of the planet, leaving them open to fresh colonization by new species, found its proper time in more recent years, when, as will be revealed, scientists have reached the startling conclusion that impacts need not always be as bad for life on Earth as the Alvarezes, Carl Sagan, and *Jurassic Park* have led us all to assume.

But the idea of such sub-lethal impacts lay dormant and unremarked in Öpik's work. Amid Cold War fever and fear about nuclear destruction raining out of the skies, cometary impact was viewed solely in terms of the megadeath. The approach of the death-star Nemesis to the Oort Cloud would, the theory went, set up gravitational perturbations that could dislodge these peaceful cosmic snowballs and send them into the inner Solar System, colliding with Earth and triggering their periodic mass extinctions.

As Dave Raup has pointed out in his wry and excellent book *The Nemesis Affair,* it could easily be that absolutely no part of this intriguing hypothesis is true, and that assessment remains as valid today as it was when he wrote it. There is no evidence for Nemesis, or indeed the Oort Cloud (beyond theory) or, beyond statistical analysis of dubious data, any 26-million-year cyclicity. Another possible nemesis for Nemesis theory is that the number of mass extinctions positively linked to extraterrestrial impacts now rests at only one (not two as Raup and Sepkoski believed in 1983), and that the end-Cretaceous extinction may not have involved a comet at all, but of course a meteorite.

Scientists are still looking for Nemesis. If they find it, it will be the biggest vindication since Alfred Wegener and continental drift. For now, however, the theory seems unable to go further. Either it's right or it's wrong. The jury awaits further evidence. The fact that it arose, and was seriously discussed in the most proper of scientific literature, demonstrates how far scientists' horizons had been widened in as little as three years from the appearance of the Alvarezes' seminal paper.

Luis Alvarez died in 1988, just before the second Cold War's end caused public concern about nuclear winter to subside. But he also missed the discovery of what for most scientists clinched the argument for an end-Cretaceous impact: a massive crater.

The lack of a credible impact site had occasionally been used to embarrass the Alvarezes' theory; but it had never really been a serious obstacle. All geologists by then appreciated that many impact structures on Earth remain unidentified. They knew that craters, if they form at all after an impact, tend to be swiftly eroded away, and that once discovered are notoriously difficult to date. Statistically, any large impact was likely to have taken place over the ocean. There was therefore a good chance that, since ocean floor is continuously consumed at destructive plate margins (like the treads of an escalator disappearing down into the bowels of the machine) any crater dating from the late Cretaceous might well by now have been lost forever.

But not long after the nuclear-winter angle was played out, the crater of doom itself was, it seemed, finally located. First discovered by oil geologists working for the Mexican oil company Petróleos Mexicanos (PEMEX) in 1978, the crater lay under the Gulf of Mexico, just offshore from the Yucatan Peninsula. Unseen at surface and visible only on maps showing local variations in the force of gravity, the circular structure was at least 106 miles across—and conveniently close to America. Journalists, editors, and NSF fund managers quickly learned how to spell Chicxulub, the name of a nearby fishing village.

Never was a science news story more exciting, better aimed, more timely, or backed by scientists and writers with higher credentials, than that which told of the great darkness that destroyed the dinosaurs—and, like all the best stories, it kept evolving, or moving on as journalists say. A funding tsunami soon flooded in, driven by the possible dinosaur connection. CGI footage showing dinosaurs, looking anxiously over their shoulders moments before being blown away by shockfronts and broiling hypercanes, became a cliché of the science documentary. The Chicxulub Crater even starred in the disaster movie *Armageddon* (1998), with Charlton Heston himself intoning a portentous voice-over, explaining that what was about to happen to the human characters in the film had happened once before.

Not surprisingly, the idea stuck. Today, among the public and scientists alike, the killer-impact theory's connection to the Chicxulub Crater is almost universally accepted as proven fact. It's something everyone knows, and everyone loves. And like so many (perhaps all) of those things that we believe in because we want to, it may still be wrong—or perhaps, as we shall see, not wholly true. Saying so, however, can make you very unpopular. Impact-induced extinction has gone from heresy to orthodoxy in less than two decades, and its connection to the Chicxulub Crater is widely regarded as proved beyond all reasonable doubt.

In establishing this new orthodoxy, pride of place must go to Luis Alvarez. Alvarez was not a disciple of Edward Teller for nothing, for both men understood instinctively how big science and big politics go together. Even those who clearly admired Alvarez and who write about him as favorably as events permit say that he "could be devastating when publicly demolishing a wrong result. Sometimes he could be personally gracious . . . and at other times not," as physicist W. Peter Trower put it, in a sympathetic memoir published by the U.S. National Academy of Sciences.

Many others, who found themselves on the receiving end of those "otherwise not" moments, speak of "brutal political attacks" and of scientific careers wrecked by Alvarez's intriguing against all those whom he saw as opponents. He was, as many physicists tend to be, a man for whom ruthless internal logic allowed little room for the slack of forgiveness. He is reported, for example, as asserting that Robert Oppenheimer (who after the war underwent grueling and public interrogations over his alleged Communist sympathies) must have been in some way corrupt. Alvarez's reasoning reportedly went thus: both he and Oppenheimer were highly intelligent men.

They could not therefore have been mistaken in either their perception or analysis of the facts. This syllogism seemed to suggest that if Oppenheimer came to hold views that were different from those of Luis Alvarez, then there must have been something else in the equation: bad faith—perhaps treason? Oppenheimer, Alvarez clearly believed, simply must have been driven by other motives.

The English poet Robert Graves parodied such reasoning in his great poem "In broken images," in which he contrasts his own slow, fuzzy and forgiving form of poetic reasoning with the logic of an unnamed "he" (widely supposed to have been the physicist Jacob Bronowski). The poem ends with the lines: "He in a greater confusion of his understanding / I in a greater understanding of my confusion." Reasoning in perfect images works well in physics. It does not work reliably everywhere, and certainly not in the judgment of other individuals and the complex processes that produce human personalities.

The ruthlessness with which Alvarez could crush his opponents went beyond scientific concerns. In June 1980, when the Alvarez paper came out, President Ronald Reagan was cutting back the space program. As the Cold War re-ignited and death from above suddenly acquired new relevance, the future funding of the threatened organizations lay in finding a new bandwagon upon which to leap. NASA's Project Spacewatch, which became Spaceguard, meshed perfectly with the idea of a rain of terror from above and the need to protect and survive, and the desire to return America to the safety and security it had once enjoyed by its very remoteness.

In battling for the acceptance of the impact theory, Luis Alvarez reserved his greatest efforts for discrediting any connection between the end-Cretaceous extinctions and the massive volcanic activity that was occurring at that time in India. The Deccan Traps are among the world's Large Igneous Provinces, as they are known, the remains of stupendous basaltic lava outpourings on a scale unknown anywhere on Earth since our species has existed. The Deccan lavas are over a mile thick, and their eroded remains today cover almost half a million square miles—though originally they would have covered about three times that area. Current and as yet unpublished work is demonstrating that the main phase of the eruption may have lasted as little as 30,000 to 40,000 years; and the difficult work of dating the lavas with meaningful precision is now suggesting that the main eruptive phase coincided exactly with the end-Cretaceous extinctions.

The idea that impact may not in fact have been the entire, central story of the end-Cretaceous mass extinctions has been the contention of a small group of researchers led by Professor Gerta Keller of Princeton University, a stratigraphic paleontologist and expert on the faunal changes at the K-T Boundary. She and her small band of co-workers have been pooping the Chicxulub party for years and receiving scant thanks.

As a proponent of the (to a physicist) messy but (to a geologist) highly plausible idea that many causes come together to create major extinction events, Keller has never denied the reality either of the Chicxulub impactor or of the collision that created the iridium anomaly. For her, both would have contributed toward making the late Cretaceous a particularly nasty time to be alive. All she has challenged—and continues to challenge—is the connection between the K-T impactor and the crater of doom. But this, for many, was the greater treason; for what she was attacking constituted Chicxulub research's publicity paydirt.

According to Keller, the Chicxulub impact happened 300,000 years too soon—and was perhaps one of a number of major (but in the Ernst Öpik sense, sub-lethal) hits that clustered around the K-T Boundary. Meanwhile, she and others supported the alternative theory, involving a number of causes based around the central cause embodied in the Deccan Traps.

Despite her distinguished credentials, Keller says that as a Chicxulub skeptic she began to find difficulty gaining a hearing at conferences or access to certain peer-reviewed journals. Her uncomfortably off-message contributions soon began to turn the subject into an apt metaphor. That there were waves goes without saying. The sulfur content of the academic atmosphere went up considerably. There was acid rain—and a good deal of noxious falling-out.

The tsunami backwash even crossed the Atlantic and lapped around my own feet for having written about Keller's work. I knew things were getting serious when, at the end of a stabbing finger, I was told one evening that I had done a disservice to science by "giving the oxygen of publicity to that bloody woman." When a fellow member of the scientific sisterhood is moved to language of that sort, you know there's something far more serious—or at least tangible—at stake than science. It was just a little whiff of the empire striking back.

The Chicxulub Crater, poster-child for the new catastrophism, with its invaluable connection to the dinosaur-killing impact, was, Keller and her coworkers came to feel, becoming immune or at least resistant to disproof. Those who believed the story, they said, closed ranks; not consciously, perhaps, but in effect—because they had now risen to control the editorial boards of certain key journals and the program committees of academic conferences.

In spite of rebuttals and deteriorating personal relationships all round, Keller persisted and in 2007 and 2008 published more critiques of the Chicxulub connection. Her studies of rocks from outcrops along the Brazos River and Cottonmouth Creek in Falls County, Texas, together with a new core (Mullinax-1, paid for by the National Science Foundation, which shows that Keller has not been exactly unsuccessful in gaining research funding for her "heresy") looked in fine detail at the stratigraphy, sedimentology, mineralogy, and geochemistry of a sequence stretching from more than a million years before the Chicxulub impact to 300,000 years after the iridium anomaly that marks the K-T Boundary. All her results suggested that Chicxulub did indeed pre-date the end of the Cretaceous by about 300,000 years. Far from being the dinosaur-killer, Keller's careful analysis of the myriad different microfossils suggested that the Chicxulub impact had no effect on living things at all.

Part of the problem in trying to do fine stratigraphy close to a major impact is the complexity of the disturbed sequences created by the massive energies released. Keller's study area in Texas, by contrast, exposed rocks of the right age that had been deposited 1,056 miles away from the impact, and therefore provided an undisturbed record of events—although her opponents claim this distance is not enough. These rocks were also relatively undeformed by any subsequent earth movements and boasted excellently preserved fossils, as well as good exposure of the sandstone complex interpreted by others as a "tsunami deposit," left behind by the wave that spread out from the impact site.

Gerta Keller had three aims: to determine the timing of the impact exactly, to search for the earliest layer containing glass beads (spherules), and to see how Chicxulub affected all organisms that lived through the experience or didn't. An impact widely supposed to have caused worldwide extinctions should, after all, have had noticeable and widespread effects on the fossil fauna.

The three cores she drilled, and the two successions logged along river beds, were sampled at 2-4-inch intervals (1-inch intervals across critical parts of the section) and samples then distributed to an international team of expert paleontologists, sedimentologists, mineralogists, and geochemists for specialized analysis. The paleontologists were asked to count the numbers of species all the way through the sequence. Others assayed trace element variations, stable isotopes, organic carbon, and studied the rocks' palaeomagnetism. The sediments were tested for bulk rock and clay mineralogy, and grain size. All spherules were analyzed geochemically using raster electron microscopy, backscatter electron imaging, and electron microprobe analysis.

The project identified the impact spherule layer in a 1.2-inch-thick yellow clay containing relict glass spherules. The clay's geochemistry matched that of altered impact spherules precisely, which was highly suggestive of a common impact origin. It lay some distance below the sandstone unit commonly held to represent the deposits of impact-generated tsunamis, and that unit itself was some way below the actual K-T Boundary as marked by the mass extinction. (No iridium anomaly was observed at the K-T Boundary or the impact spherule layer in any of the Brazos sections, though up to three small anomalies are present within and slightly above the sandstone.)

Keller's investigation suggested that before the impact occurred, quiet, shallow marine sedimentation was taking place in late Cretaceous Texas, at depths in the ocean of 87–109 yards in a warm climate. Mudstones and claystones with burrows bore abundant evidence of a rich diversity of life. Then the Chicxulub impactor hit Yucatan, Mexico, 1,056 miles away.

Within hours or days of the impact, a thin layer of impact melt glass spherules rained down and covered an area from New Jersey to Brazil. Although tsunamis certainly swept outward from the impact site, their deposits were subsequently redistributed by prevailing current regimes because Keller and her team found no evidence of any significant disturbance above or below the spherule layer either in Texas or in northeast Mexico.

After the impact, things returned to normal; but over the next 200,000 years the global sea level fell—by about 66 yards in all. Nearshore areas were exposed to erosion. In the shallow-water environment, currents scoured small canyons and valleys, into which eroded sediments from nearshore areas were dumped. The first sediments to arrive included many rip-up clasts of already lithified sediments that themselves contained impact spherules. Some clasts displayed desiccation cracks in which spherules had become trapped, showing that the impact layer had been exposed to erosion.

Toward the top of the sandstone unit, hitherto interpreted as the tsunami deposit, conditions began to deepen once again, and layers of truncating burrows show the repeated recolonization of the seabed over an extended period. At the top of the unit, normal marine deposition returned with the deposition of an upward-fining calcareous silty mudstone that grades into a dark organic-rich claystone. The water was poorly oxygenated, with a low-diversity microfossil population that prevailed until the K-T Boundary itself. The sandstone unit, in other words, post-dated the Chicxulub impact by a long time

and far from being a Tuesday-afternoon deposit, laid down
in a few glorious hours, actually represented a very extended
period of perhaps tens of thousands of years.

The case is still hotly debated. Before Keller's most recent
papers were published, during the winter of 2003–4, I had the
pleasure of acting as ringmaster to an online debate between her
and Professor Jan Smit of the Vrije Universitet, Amsterdam—
the man whom Walter Alvarez regards as co-discoverer of the
K-T anomaly. Smit and Keller slugged it out on the website of
the Geological Society of London, where the debate can still
be read, over several weeks. It was an interesting lesson in scientific controversy because—then as now, it seems—when all
else was cut away, the argument came down to the interpretation of particular disputed objects. For one side, a forami-
niferan fossil is unequivocal evidence of date. For the other
side, it is nothing of the kind because it has been reworked
from an older horizon and re-deposited in younger sediment.
Or it is not a foram at all, but a collection of dolomite crystals that just look like one. When you reach this point, further
debate becomes futile; there isn't anywhere left to go except
back into the field. This is what Keller did, and the results she
has obtained since have reinforced her conviction—though of
course they have done little to convince Jan Smit.

Keller's opponents flatly disbelieve her claim that she has
suffered unfair discrimination for her off-message views and
point to her continuing success at obtaining research funds.
They too, they say, suffered their share of rejected papers and
find it hard to credit Keller's assertion that she has found publication difficult. The rebuttal of her recent Brazos River work
in a 2008 reply by Jan Smit and six colleagues came down as
usual to disputed objects. What Keller and supporters see as

the impact spherule layer 300,000 years too low in the sequence, Smit and colleagues say is nothing but volcanic ash. Layers that Smit et al. believe were deposited suddenly by tsunami, Keller and associates say are full of burrowed horizons that show it to be untrue, and so on. Attempts to bring in third parties to examine disputed evidence have not been successful, collapsing amid mutual recriminations and charges of obstruction. Gerta Keller is now pursuing work in the Deccan Traps of India, while Smit no longer works directly on the problem. Sadly, the two no longer engage in scientific debate of any kind. One senses in both a deep feeling of fatigue.

Keller says: "Conventional wisdom holds that any such large impact leaving a 109-mile-diameter crater would cause major mass extinctions. But this hypothesis is based solely upon the assumption that Chicxulub was the K-T killer. No other major mass extinction in Earth history is associated with major impacts. This hypothesis has no empirical support and must be considered false—at least with respect to Chicxulub."

A bold statement; but Keller thinks there has been a sea change in the Chicxulub controversy. Since the days when she found it difficult to get a hearing for her views, she tells me, it has become much easier to publish in high-profile journals. She feels that fewer of the key influential positions are now held by proponents of the impact theory, which she feels has rather stagnated while its opponents seem to be producing the most significant new data. "I think this impact theory is slowly dying," she says, though she admits this may just be wishful thinking.

Many other scientists talk of the "full Alvarez" (crater and all) in language that sometimes lapses into the tropes of popular religion—"the greatest scientific story ever told"—and even compare Keller's continued doubts about the new gospel with laughable assaults by religious literalists on such well-established facts as organic evolution and the age of the Earth. But Keller continues to spit in church. "Tying the K-T mass extinction to volcanism has had a large following among vol-canologists, but the age-control [on the eruptions in the Deccan of India] were lacking. We have [now] been able to do that, and are gathering data to find out just how volcanism did its deed."

As I write this, Keller is writing a book chapter describing how the Chicxulub connection with the K-T impactor was finally debunked; while at the same time, those who cleave to the full Alvarez are busy rehashing old results in multi-author review papers that barely mention Keller's work, and seem to call on sheer force of numbers—as though democracy mattered in science. If that were true, none of the falsehoods that everyone believed for centuries would ever have been disproved. Indeed, if science history teaches anything, it is that—as Luis Alvarez knew well, because he and his team had to do it in 1980—far from being welcomed, all truly original and revolutionary ideas have to be rammed down the throats of the disapproving chorus that nearly always greets them.

Science's much-vaunted self-correcting mechanisms can go wrong sometimes, and the review of one's work by peers can easily turn into the oppressive rule of gang culture. Peer review, the process that makes scientific publishing and funding "scientific," is, like democracy, a best worst form of government. No one in his right mind would seriously suggest ditching it just because it has weaknesses. Eternal vigilance is the price of liberty, and freedom of speech the first casualty when we fail to exercise it. As in politics, so in life; and science is another part of life—something that people do.

＊

When G. K. Gilbert argued himself into error over Meteor Crater, he had to make the meteorite fragments that surround it a coincidence, which complicated his cryptovolcanic explosion theory beyond the point where he began to feel the nick of Occam's razor. In that instance, Occam was giving good advice; but one can become over-fond of simplicity. Physicists, in their subject, are rarely misled by the instinct; but geologists, dealing with complex processes and contingent events spread out over vast periods, are dealing with history, not physical principles. Luis Alvarez, looking for generally applicable laws, would have been disappointed to discover that, instead of finding a universal rule about mass extinction, such as Nemesis, he had maybe only discovered an event—something that happened just once, without precedent, and which expressed no bigger scientific rule than shit happens.

Geologists are more ready to mistrust overarching laws and single-cause mechanisms. For them it is much more likely that mass extinctions have not one but many contributory causes and that the severity of the event depends on how many of those causes happen at once. Derek Ager, writing in 1976, imagined a multitude of natural cycles, all doing their own things and remaining mostly out of phase with one another until the unhappy time arrives when, quite by accident, all suddenly swing the same way together. Then, as Gerta Keller put it with regard to the K-T extinction, the Earth experiences "a progressive multi-event catastrophe—a concerted assault on the whole edifice of life by a combination of massive volcanism, multiple impacts, and their associated climatic and environmental changes."

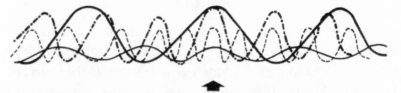

Derek Ager's "silly diagram." The arrow shows where various environmental cycles suddenly and accidentally come into phase, so joining forces to produce a potentially catastrophic effect.

Derek Ager was mathematically literate enough to call this his "silly diagram"—a graph without either axes or values. But despite its mathematical meaninglessness, it encapsulates what Ager saw as a profound truth about the way the Earth works, and which had been fermenting in his mind even as he and I were together on Gotland and Per Thorslund on the mainland was four months from rediscovering his specimen of a fossil meteorite.

According to Ager's model, the end-Cretaceous extinctions would have happened anyway without an impact, albeit less severely, simply as a result of global regression of the sea and truly colossal volcanism. During the Mesozoic Era's last few hundred thousand years, a shower of massive impactors culminating, just as the Deccan eruptions reached their peak, in one last impact that created the worldwide iridium anomaly was simply the straw that broke *T. rex*'s back.

If science can be said to do one thing for the rest of us, it is to make life more comfortable physically but less comfortable psychologically. Sigmund Freud usually gets the credit for being the first to notice this and expressed it in terms of science dethroning humans from the center of things. First, Earth was banished from the center of the universe. Then humans became a product of evolution from brute ancestors, and so on. But Darwinian evolution does leave us one shred of self-esteem; namely that we can still think of new species, and therefore ourselves, as in some way better than those we replaced. Darwin himself wrote in the *Origin:* "If . . . Eocene inhabitants . . . were put into competition with existing inhabitants, the Eocene fauna or flora would certainly be beaten and exterminated." Human beings may not have been a special creation, but by virtue of natural selection they were better. It made everyone feel a bit like aristocracy.

But mass extinction, especially if visited upon us by random cannons in the great playerless pool-game of the Solar System, kicks away even that feeble prop to our vanity. No amount of natural selection can fit an animal to survive a 6-mile space rock landing on its head. In between such moments of terror, during the intervening ages, Darwinian processes prevail. But if the story of life could be said to have a plot, its major turning points were catastrophes arising by chance.

Because I have written about these debates as an observer, I am often asked what I think about it all. Did the Alvarez impact hypothesis really lead scientists to the smoking gun of the end-Cretaceous impact at Chicxulub, so conveniently close to America, where most of the interest and money was centered? Or did "seek and ye shall find" just come true, the way it usually does?

When the debates have gone on to exhaustion, where all one has left are the disputed objects, scientists must go back to the field and the rest of us—observing planet science as though from orbit—have to go with our instincts. Many people whose

scientific judgment I cannot fault think Keller is at best mis-
guided about the Chicxulub crater and, in a curious echo of
Luis Alvarez's opinion of Robert Oppenheimer, conclude she
must be driven by unspecified other motives.

The implication, one feels, is that Gerta Keller should be
ranked among the denialists, that species of contrarian that
persists in a belief despite all current evidence, for underlying
reasons of their own. Young Earth Creationists are in denial
over the age of the Earth and evolution because they prefer
to hold to a literal interpretation of their creed. AIDS denialists hold that the condition is not caused by the HIV virus,
and many do so because they themselves are HIV positive or
have their own commercial reason for peddling useless herbal
alternative therapies. All denialism is rooted in self-interest
and conspiracy theory and pretends there is a debate going
on when there is none. Those who say the Holocaust never
happened pretend there is disagreement among historians for
the mass murder of Jews by the Nazis. Those who deny global
warming pretend that there is serious disagreement and mali-
cious cover-up among leading climate scientists. Some believe
that the truth of the collapse of New York's World Trade Center
Twin Towers in the 9/11 attacks was a financially motivated
scam. All these people use the tactic of stoking up debate over
issues that are long settled.

Galileo said that the authority of thousands counts for
nothing before the single voice speaking truth. The annals of
the Nobel Prize bulge with the names of scientists whose work
was almost stifled by the opposition of their peers because it
challenged orthodoxy. Yet for every would-be lone genius there
are a thousand pig-headed contrarians, and many acknowledged mold-breakers went on from their great breakthroughs

to challenge orthodoxies elsewhere in science and failed disastrously—never quite recapturing the magic of that one occasion when their sheer inborn cussedness paid off.

Personally, and in common with many journalists, I like doggedly difficult people—and not simply because their relative loneliness has the romantic allure that makes good copy. I like the awkward squad, irrespective of whether I think they may be right, for it is not an observer's place to pass judgment. My only wish is to paint a picture of a scientific debate, when the stakes are high and the workaday safeguards of the scientific method are put to their sternest test. After Keller and Smit withdrew from the Geological Society's online debate— both bloodied, exhausted, but unbowed, still clinging firmly to their own interpretations of the disputed objects—I had one over-riding reflection. It involved a story from geology's allied trade of archaeology. It is not a valid argument in science; but I offer it anyway, remembering that the great physicist Ernest Rutherford was wrong about nuclear energy and the novelist H. G. Wells was right.

The tale begins with the German archaeologist Heinrich Schliemann and the first works of Western literature—the *Iliad* and *Odyssey*. Few stories can rival these tales for significance in European culture. They are the founding texts of the West, and they exerted a powerful influence over Schliemann from, according to him, the age of eight, when he resolved to be the first man to excavate Troy. For Schliemann believed that the places and events mentioned in these great mythic poems were both geographically and historically real. In the *Iliad*, Homer depicts the famous siege of Troy by Greek forces—including, of course, Odysseus—under their king, Agamemnon. Agamemnon, head of the doomed house of Atreus, suffers a miserable fate after the war is over when, returning victorious to Mycenae, he falls victim to his treacherous wife Clytemnestra and her lover Aegisthus.

Hoping to discover the tombs of these mythical characters at Mycenae, Schliemann was excited to discover a large cemetery outside the walls of the acropolis, within which were many shaft graves arranged in a number of circles. Grave circle A proved to be one of the richest archaeological finds ever made. And within Grave V, Schliemann came upon three male burials.

Two of the deceased wore gold death-masks, one of which Schliemann identified and named as the Death Mask of Agamemnon. He telegraphed Greece's King George I on November 16, 1876: "Your Majesty, it is with great pleasure that I inform you that I have discovered the tombs which, according to Pausanias' account, belong to Agamemnon, Kassandra . . . who were murdered by Clytemnestra and her paramour . . ." The seeker had found, as seekers are promised they will. And, as seekers will, he had found what he wanted to find. He had found what his ruling theory, the Homeric myth and the historical account of Pausanias, demanded.

The Mask of Agamemnon, its eyes shut, now sleeps in its glass case at the Greek National Archaeological Museum in Athens. Still known universally by the name Schliemann gave it, the mask's true nature has exercised scholars ever since—it is perhaps archaeology's ultimate disputed object. To some, a minority, it is a forgery, or an artifact from elsewhere, salted by Schliemann among the finds of Grave V to prop up his theories and boost his reputation. Schliemann was not above the odd piece of double-dealing and is known to have been something of a romancer. But the truth is probably less exciting. The

Mask of Agamemnon cannot be what Schliemann thought it was, simply because it is too old, pre-dating the Trojan War by four centuries. If cuckolded King Agamemnon was ever buried with a golden death mask, both it and his remains lie elsewhere—or were destroyed long ago.

✳

The geologist and science historian William Glen wrote at length about the plate tectonic revolution and found, to nobody else's very great surprise, that all the participants he interviewed after the event edited their recollections so as to put themselves in the best possible retrospective light. Glen therefore seized upon the Alvarez theory as a scientific revolution that he could study as it unfolded.

While still in the middle of the project, he "surprised the hell" out of Stephen Jay Gould one day by telling him that he believed the significance of the Alvarez impact theory would come to surpass that of plate tectonics. This sounded to Gould like an extraordinary heresy. Plate tectonics is the grand unifying theory of geology and has left no branch of the science untouched. But Gould had misunderstood. Glen had perceived that, unlike plate tectonics, which really only affected scientists, the Alvarez impact idea had changed everyone.

By the mid '80s, hardly an eight-year-old in the U.S. did not know that life on Earth was once almost wiped out by a meteorite, and could be again. Not an adult anywhere could be excused ignorance of the possibility that we ourselves could bring life on Earth to the brink of extinction in nuclear war by a similar mechanism. The H-bomb, which Luis Alvarez helped to bring into being, had now become so terrifying that the prospect of its use worried even the most deluded survivalist.

Feature films crowned the idea's graduation from scientific theory to popular myth. Plate tectonics was an arcane sideshow by comparison.

In 1991, an international working group under the auspices of NASA convened to conduct a survey of asteroids whose orbits bring them inside 1.3 astronomical units of the Sun. The report gave rise to what became Spaceguard Survey, a name borrowed from Arthur C. Clarke's novel *Rendezvous with Rama* (1973), which opens with an account of an impact upon Earth in 2077. The idea of NASA's Spaceguard project soon spread across the world. In the UK, astronomer Ernst Öpik's grandson Lembit was sitting in the House of Commons. He understood, from his grandfather's work, the threat posed by impacts. Following an evening lecture at the Shrewsbury Astronomical Society on September 25, 1998, Öpik promised to bring the matter up in the House of Commons—which he did, on March 3 the following year. The government of the time duly established a task force, which produced a report on Near Earth Objects in September 2000. By late February 2001 all of its recommendations were accepted.

Meteorites, and the threat they could pose to civilization, had hit the mainstream to the extent that even politicians had noticed. Together, Alvarez father and son had changed people's view of their world, of its place in the cosmos, and of humanity's place upon it. A whole generation now lived with new knowledge—and faced the terrifying prospect that one day, all that might stand between them and the fate of the dinosaurs could be some fool on a hill with a telescope and a rocket-bomb hurtling through space—Bruce Willis at the helm.

For centuries everyone had assumed that the best by-product of scientific endeavor was to banish fear and superstition. Now, the Alvarezes—and the authors of over 2,000 related papers that were published in the 1980s alone—had seemingly brought the Enlightenment itself to an end under a nuclear pall as dense as that which they envisaged snuffing out the Middle Earth of dinosaur and ammonite.

Had science freed us from old, irrational fears, only to replace them with new, rational ones? It seemed so. Children could now be frightened into compliance by powers more mighty than dinosaurs or sea monsters; the most ancient of ancient menaces, stalking the silent voids of space, darker than night, older than sunlight, waiting to strike the Earth goddess Gaia to the heart and leave nothing of her but fire, dust, bitter cold, and death.

As geologists re-embraced the narrative of catastrophe, the world welcomed back the poetry of demons.

PART THREE

DELIVERANCE

7

LIFE IS EVERYWHERE

In 1969, the United States passed a piece of legislation into the Code of Federal Regulations that became known, perhaps jocularly, as the Extra-Terrestrial Exposure Law. Although never invoked in court, it did have one major consequence, which those of us old enough to remember the first Moon landings may recall. The first astronauts to visit the Moon had to spend weeks in quarantine after they came back, just in case they had picked up something nasty while they were out there.

This inconvenient practice was discontinued after Apollo 14, but the idea that outer space might be alive with opportunistic alien bugs still haunts the popular imagination, and pervades science fiction from 1950s B-movies to the present day. The history of life has probably been influenced many times by meteorites, hastening extinction in one case at least, but also stimulating evolution in another—as we shall see. The idea that meteorites might promote evolution in some way is, once again, not new. However, early notions envisaged a process akin to infection; suggesting that alien biological material pervaded the cosmos and, carried by meteorites, rained down upon us all the time. So while it afforded meteorites a more positive, life-bringing image, the envisaged mechanism nevertheless still carried an implied and sinister threat.

It all sounds distinctly wacky today; but the possibility that life or its traces might be found in space was not always a sideshow. In fact it once, briefly, held center stage in one of the most pressing scientific debates of an age—namely, the question of whether life on Earth was created just once in the dim geological past or whether it is generated continuously, all around us, all the time.

✳

At ten minutes past eight o'clock on the eve of Pentecost, May 14, 1864, above a string of small villages stretching south and east of Montauban, France, a white fireball streaked across the sky out of the northwest at 12 miles per second and exploded "like the bouquet of a firework"—in the words of "Peyridieu," writing in the May 17 edition of *Le Courrier de Tarn-et-Garonne*.

After the explosion, at an altitude of about 12 miles, the fireball became a dull red as cooling fragments were strewn across an area covering about 3 square miles. At the point of the explosion, a white cloud hung in the high atmosphere, as sounds like a distant cannonade echoed for several minutes from horizon to horizon. The largest piece recovered, of the twenty or so known to have reached the ground, was about as big as a football and landed just outside the village of Orgueil.

The Orgueil meteorite was of a very unusual sort; less like the usual hard, stony or nickel-iron object, and more a great, flying black pudding. It was a carbonaceous chondrite, a class that makes up less than 4 percent of all known chondrite meteorites. Even today, only 560 are known, yet they have been divided into no fewer than seven sub-subcategories. Orgueil was an example of the rarest category of all—dubbed "CI," for soft, black objects, paradoxically containing no actual chondrules at all. CI meteorites have a chemical spectrum that resembles that of the Sun more closely than any other meteorite. For this reason many scientists believe that CI meteorites are the remains of extinct comets that, having passed repeatedly through inner space, had been dried of their volatiles under the heat of the Sun.

You would expect a gooey meteorite to disintegrate quickly, and for that reason the CIs known to science today (a mere five) come from observed falls. Their most common component is magnetite, a black, magnetic iron ore, mixed with various carbonate, sulfate, and sulfide minerals. The Orgueil meteorite was especially rich in magnesium sulfate—Epsom salt, visible as white specks and veins, and like others in its class it was porous, rich in water and carbon compounds, and low in density as a result. The carbon molecules it contained formed long chains, chains we traditionally associate with the chemistry of life—and to which we give the name (with typical biocentrism) organic.

As with the falls toward the end of the eighteenth century, the arrival of this meteorite could not have been timelier in the history of science, nor better placed—for France was then at the center of a huge debate over the nature of life itself. How did life arise? Did non-living matter possess a vital impulse that could create flies and mold and maggots by spontaneous generation, as many had held since ancient times, or did every living thing have to come from some sort of seed or egg laid by an ancestor in the same line, like chickens? The arrival of a meteorite apparently containing chemicals associated with living processes placed the object center stage in this highly charged biological battle then raging. It also linked together the ancient idea of spontaneous generation with the idea that if non-living matter possessed the vital spark, life could then be expected to pervade the entire cosmos, an idea known as panspermia.

The battle over spontaneous generation was raging among, on one hand, the greatest microbiological pioneer of all, Louis Pasteur, and on the other Félix Pouchet, director of the Rouen Natural History Museum, and a newly qualified medical doctor and political activist, Georges Clemenceau. It is not widely appreciated today that Clemenceau the renowned French statesman was also an indefatigable journalist, whose assembled articles would one day fill ten 320-page volumes. Unlike Clemenceau, the scientist Pasteur was not medically qualified; but as well as attacking his competence, Clemenceau, a lapsed Protestant and fervent secularist, smeared Pasteur's opposition to spontaneous generation by suggesting that he was driven by the need to affirm the Catholic doctrine that life was breathed into matter on the third day when, according to Genesis 1:11, God said "let the Earth bring forth grass."

Pasteur was indeed a devout Catholic, but whether his conviction was driven by religious instinct hardly matters; the great biologist was simply right. Life could only come from other life—and he proved it in a series of dramatic and breathtakingly simple experiments. The crucial test, carried out between 1859 and 1865, involved boiling broth in a swan-necked flask to sterilize it. The narrow S-shaped tube leading from the mouth of the flask swept up, then down, ending in a final upward turn that cut off the flask's contents from all airborne micro-organisms—though not from the air itself. Once rendered sterile, the broth should have kept perfectly; and

Pasteur showed that this was indeed so unless the swan neck was broken, or the flask tipped so that its previously sterile contents came into contact with the bottom of the trap where airborne bacteria and yeasts had settled.

But although spontaneous generation was conclusively debunked by Pasteur, the controversy rumbled on for years—its adherents attacking Pasteur's methods and devising experiments in vain attempts to refute them. Pouchet went to his grave still believing in it; though by that time others, including Clemenceau, had conceded defeat. As the history of the related idea that life pervades the cosmos—panspermia—shows, for those who truly want to believe, evidence or lack of it is not necessarily an obstacle. The truth is always out there; it just may not be the truth you want.

Pasteur delivered the *coup de grâce* to spontaneous generation in a lecture to the Sorbonne on April 7, 1864, and one month later the Orgueil meteorite claimed the great microbiologist's attention. But he was not the earliest on the scene. The honor of being the first scientist to tackle the Orgueil meteorite went to François Clöez, professor of chemistry at the Paris Muséum d'Histoire Naturelle. He reported organic matter, which contained the elements carbon, hydrogen, and oxygen and resembled humus or brown coal (lignite). The Orgueil meteorite in fact contains just over 3 percent carbon, all of whose atoms are bound up in complex, organic molecules. Clöez believed this pointed to the "existence of organized substances in celestial bodies."

Two years later, in 1866, chemist Pierre Berthelot, the man credited with discovering that all living things contain carbon, hydrogen, oxygen, and nitrogen, detected hydrocarbons using a more sophisticated analytical method. Berthelot never subscribed to the notion that this indicated some sort of exobiological origin, but the connection between the Orgueil meteorite and the chemistry of life grew stronger. Allegations of accidental contamination were never enough to explain entirely the presence of complex carbon molecules, and soon the great Pasteur was drawn in. Perhaps the meteorite contained actual cells? To test for this, Pasteur used a drill, which he hoped would remove samples from inside the meteorite without introducing contaminants, and attempted to culture any micro-organisms. Results were negative. Interest faded but not everywhere. Popular science, especially where it grades into speculative fiction, remained fascinated by bugs from space.

Browsers in French bookshops will be familiar with the name Flammarion, the publishing house founded by Ernest Flammarion in 1876. Ernest's brother Camille was an astronomer and writer who in 1864, the year that Orgueil fell, included a long new section about meteorites in his popular book *La Pluralité des Mondes Habités* ("On the plurality of inhabited worlds"). In this, Flammarion opined that the presence in carbon-rich meteorites of carbon, oxygen, and nitrogen clearly indicated that life existed throughout the cosmos. Flammarion was France's version of England's H. G. Wells, and as a result nobody in the scientific world took much notice. But things took a different turn in the 1870s when two eminent physicists, Hermann von Helmholz in Germany and William Thomson, 1st Baron Kelvin, in England, apparently independently revived the idea that comets and meteors might spread the germs of life through the universe.

By 1871, science had become aware of only seven carbonaceous meteorites in total. Nevertheless, Kelvin, who believed strongly in the plurality of inhabited worlds, felt able to suggest that, thanks to innumerable and inevitable collisions between celestial bodies, it was "probable in the highest degree" that there existed "countless seed-bearing meteoric stones moving about through space." The idea that life on Earth might have been seeded from such "moss-grown fragments from the ruins of another world" might seem "wild and visionary," but was not "unscientific," he said, concluding his presidential address to the British Association for the Advancement of Science in Edinburgh.

Today, Kelvin's preference for intelligent design over evolution and his beliefs that the Earth could not be older than 20 million years, that radio had no future, X-rays were a hoax, airplanes were impossible and there was nothing new to be discovered in physics, all conspire with hindsight to take the shine off his grandeur. Yet Victorian society greeted his many pronouncements with only slightly less reverence than those of an archbishop, so his public avowal of panspermia served to keep the idea before the scientific community. Clemenceau might have suspected Pasteur of harboring other motives, and possibly religious ones, for denying spontaneous generation. But there was no doubting Kelvin's motive for favoring panspermia, since he made no secret of it. Panspermia was the great authoritarian's recipe for putting God back into creation, and opposing the materialistic idea—implied by Darwin in the *Origin of Species*—that life arose just once, from non-living matter, way back in the distant history of the Earth.

There is a fairly consistent pattern about interest in panspermia, which itself is a broad church. At one end are those who merely acknowledge that organic molecules in meteorites and comets may have seeded the Earth and influenced our biochemistry. At the other end are those who envisage actual living things floating through the cosmos. But however you define it, there is nothing inherently unscientific about it as an idea; it's just that as far as bugs in space go, there is no convincing evidence. Panspermia never goes very far beyond the stage of being an intriguing idea; but, like crime and poverty, it never goes away completely. Significantly, as with Kelvin, it tends to lurk in the minds of those who, for whatever reason, find it hard to accept Darwinian evolution.

Scientists tend to judge ideas, perhaps unfairly, by the company that they keep. Thomas Huxley, Darwin's doughty public defender, had no time for panspermia. For him, it was "creation by cockshy—God Almighty sitting like an idle boy, shying aerolites (with germs), mostly missing, but sometimes hitting a planet!" As the *New York Times* might have put it, Baron Kelvin appeared to Huxley to have exchanged one mystery for another.

✳

"Orgueil" means pride, and the question soon arose as to how the pride of Orgueil could be preserved for posterity. Being soft, friable, and wet, the meteorite posed greater than usual problems for curators. In the Muséum d'Histoire Naturelle, a fragment weighing 4 pounds was placed in an ice box under dried air. At the Montauban museum, closest to the site of the fall, two smaller specimens were sealed up tightly in glass cases. These measures helped preserve the fragments but at the same time rendered it most inconvenient to carry out research on them. Perhaps this was why, despite all the excitement of finding organic carbon in outer space, no further work was performed on the Orgueil meteorites until almost a century later.

Despite Pasteur's failure to find it, the idea of meteoritic life resurfaced in 1932 when Charles Lipmann, a bacteriologist from Berkeley, California, took sixteen specimens of various meteorites and successfully cultured rod- and sphere-shaped bacterial cells. His work was picked up by the *New York Times* and created a flurry of excitement; though others soon cast doubt on Lipmann's results, suspecting laboratory contamination. The *Times* reviewed the subject again four years later, but by then even Lipmann had backed off. Bugs in space became dormant.

As the space age got properly underway, they sprang to life again. In 1961, Bart Nagy and Douglas Hennessy of Fordham University, New York, and Warren Meinschein of oil giant Esso analyzed fragments of the Orgueil meteorite using a mass spectrometer, which separates molecules of differing weight. They found paraffinoid hydrocarbons (organic molecules consisting of simple, straight chains of carbon atoms). These authors, catching the mood of the hour, courageously concluded that "biogenic processes occur and living forms exist in regions of the universe beyond the Earth." The *New York Times* quoted them as saying: "We believe that wherever this meteorite originated something lived." Less than a year later, in 1962, Nagy and George Claus reported even more complex organic molecules from the Orgueil meteorite and the similar Ivuna meteorite from Tanzania—including amino acids, the building blocks of proteins. They also reported what they called organized elements—that turned out to be common mineral structures.

This was all very exciting, but its main scientific consequence was the Extra-Terrestrial Exposure Law. Its second consequence was that suddenly museums all over the world found themselves being begged to send analytical samples of any fragments they might possess of the famous Orgueil meteorite. One such museum was that of Montauban.

One of Montauban's two long-sealed specimens traveled to Chicago, while the other went to Bart Nagy in New York. The first fresh eyes to look closely at these specimens for ninety-seven years discovered something amazing embedded within the body of the meteorite. These were not the hoped-for space bugs, but plant fragments and a piece of coal. The plant was soon identified as the common reed *Juncus conglomeratus*. At first, accidental contamination was suspected, but this was ruled out when closer inspection showed that shards of the meteorite's original fusion crust had somehow become buried within its body. Then it was realized that part of the specimen's crusted surface had been sealed with glue to make it resemble the crust that had been broken. This was not a simple case of contamination. This specimen had been doctored.

Who conceived this practical joke will probably never be known. Even the culprit's motive is hard to fathom. The temptation as always is to build an interesting story involving some plot, aimed perhaps at discrediting the high-profile scientists involved in the great controversy over spontaneous generation. Yet Pasteur never went anywhere near the Montauban specimens and would hardly have been fooled if he had. The whole case has more the air of a prank about it than a true hoax, recalling the story of the unfortunate Dr. Johann Beringer of Würzburg, in Bavaria, of whom we shall have more to say in the next chapter. Pranksters planted carved stones for him to find. Completely taken in and believing them to be natural objects, the hapless academic lovingly described them in a large illustrated book and was ruined by the shame of it all. Beringer was targeted because he had made enemies with his arrogant condescension. Perhaps at Montauban, some disgruntled employee with access to specimens tampered with the evidence in the hope that he might make a fool of his superiors. Whatever the reason behind the hoax, the bait was never taken. Now, in all probability, the tawdry truth about how the pride of Orgueil became prejudiced has gone forever to the grave.

So, assuming that life did not come to Earth aboard meteorites as bits of reed embedded in a flying black pudding, how might it actually do so? Conclusive proof that organic chemistry was not the sole preserve of living things came in 1953 when Stanley Miller published the results of a truly great experiment designed with his research supervisor, the Nobel Prize–winning cosmo-chemist Harold Urey. Assuming that the atmosphere of the early Earth would have contained a noxious mix of ammonia, methane, and hydrogen, Miller and Urey sealed some in a chemical apparatus through which an electric spark was passed, to simulate lightning. They let the experiment run for a week, after which as much as 15 percent of all the carbon in the mixture had combined in complex organic compounds, including amino acids. While this suggested that the chemical precursors of life could have arisen spontaneously on the early relatively oxygen-free Earth, it also suggested that such processes could also occur in oxygen-free space. Meanwhile, astronomers have since discovered that if space can correctly be said to be full of anything, it is full of organic molecules.

But there is one crucial distinction between the organic chemistry of life and that of the test tube. One of Louis Pasteur's earliest discoveries was that complex organic molecules can come in right- and left-handed forms. He had noticed that solutions of tartaric acid derived from purely natural sources (wine, vinegar, and so on) altered the polarization of light passing through them. Contrastingly, tartaric acid solutions synthesized in the laboratory had no such effect. Pasteur discovered that molecules of tartaric acid could come in two forms, each a mirror image of the other—a fact mirrored in the solid crystals that they formed when dried. Truly organic tartaric acid came in one form only, whereas the synthetic version produced both. After evaporating the solution of synthetic tartaric acid to dryness, the left- and right-handed molecular forms could be separated by hand using a fine paintbrush and a magnifier. If these separated crystals were then re-dissolved, the solutions now polarized the light, just like natural ones.

Thus, when biochemical processes synthesize molecules they favor one molecular "handedness" over another. A predominance of left-handed amino acids or right-handed sugar molecules can therefore be taken to indicate biological influence. The amino acids found in carbonaceous meteorites show little sign of this, and if they do it is relatively slight and usually attributable to contamination.

Recent work on pristine carbonaceous chondrite meteorite samples collected under sterile conditions in Antarctica, where organic contamination is likely to be minimal, poses an interesting problem. Although the scientists who discovered a slight imbalance of left- and right-handed amino acids fell short of advocating bugs in space, they did suggest that life's peculiar preferences in amino acid stereochemistry might have originated when the Earth was long ago seeded with molecules that predominantly dressed to the left. Although this case is far from closed, the overwhelming probability remains that meteoritic organic molecules are organic in name only.

✳

The idea that space was alive remained unproved. It is often said that remarkable claims require remarkable proofs, and that is just as true of panspermia as it is about the theory that a meteorite killed off dinosaurs. Both theories require scientists to stray widely from their core disciplines. But the success of the Alvarez hypothesis, where a physicist came into geology, provides an instructive contrast with the abject and embarrassing failure of a cosmologist to enter biology. For the most recent attempt to put panspermia back into mainstream science was spearheaded by one of the greatest scientists of the twentieth century, Professor Sir Fred Hoyle.

Most scientists agree that crossing disciplinary boundaries is generally a good thing; but it takes extreme care to do it properly. When hapless tourists wander accidentally into tough parts of unfamiliar cities, the frequent result is that they get mugged. Much the same thing can happen to scientists who practice "the genteel art of academic trespassing," in Walter Alvarez's memorable words. The correct approach is to employ a local guide. No one would say that Luis Alvarez couldn't look after himself in a fight, but he could not have prevailed without the assistance of his geologist son. Woe betide any scientist who tries to tackle strange subjects without the diplomatic and intellectual help of a native sherpa.

The danger of embarrassment and humiliation increases with the status of the interloper—and not just because he has further to fall from grace. If an academic trespasser, suffering from some elementary misconception, submits his naïve idea to peer review, then he will be picked up by the border police before the trespass even begins. This will also apply (or at least it should) to trespassers of great repute in other fields, for when it works properly the mechanisms of science are no respecters of nobility. However, peer review is necessarily conservative; and it may, and occasionally does, reject unfamiliar, revolutionary ideas unfairly. Researchers with high reputations are then at their most vulnerable. Because they may be vain enough to have difficulty believing that they could actually be mistaken, they may decide to bypass peer review. Rather than grasp that they are being saved from themselves, they see a closed shop, acting as a barrier to independent thought.

I remember once sitting as a schoolboy during one of those hot, post-exam weeks at the end of summer term, in a period of personal study. As the rest of us fifteen-year-olds passed notes and chewed gum, a studious classmate quietly busied himself in a corner with a pencil and a four-figure logbook. Toward the end of the period, being supervised by a member of the physics department, this guy had convinced himself that he had proved Albert Einstein wrong on a key point of relativity. Not one to underestimate his own abilities, he decided to show his work to the master. He had been unwise enough to share his excitement with us earlier in the playground. His notorious self-importance had for years made him a figure of mockery, so I fear we may have encouraged him in the hope that something good might happen later to relieve our boredom. We were not disappointed when after laying his workings before the head of physics, who was busy filling in report books and resented the intrusion, he was dismissed with scorn and left to walk ignominiously back to hoots of ignorant derision from us.

When I see this exact same thing happening to eminent adults on an even more public scale, I find myself wishing that they had suffered a few more humbling experiences of this sort when they were young. Alas, the determined and over-confident trespasser intent on bypassing peer review passes easily through the next safety net, which should be provided by the near impossibility of publishing anything. Knowing that big names sell and careless of much else, book publishers are only too flattered to add them to their lists, and so manuscripts that would normally be met with polite letters of rejection are published. Such books challenge accepted wisdom (for which read: the opinion of scientists who know what they are talking about) and so find gaining media attention easy—especially with the added stardust of great names. Heresy is news; orthodoxy is not. Journalists do not have to apologize for this, though many scientists—confused perhaps about the difference between truth and news—think they should.

Revolutionary ideas like those of the Alvarezes stimulate fruitful research even from those who initially resent the intrusion because they are based on sound reasoning and new data. The trespasses of the ill-informed, because they can be easily rebutted by simple recourse to the basics of a subject, with which the illustrious invaders did not condescend to acquaint themselves, are simply boring. Yet, because of media brouhaha, there can be no avoiding them.

By 1985 I had just ceased to be a practicing scientist and was now writing about it instead. To my joy, *New Scientist* asked me to write a piece about a controversy that had blown up over a startling allegation of fossil fraud being made by Britain's foremost scientific knight-errant, the great Sir Fred Hoyle, the genius behind stellar nucleosynthesis, together with fellow astronomer Chandra Wickramasinghe. This fraud had been no mere prank, they alleged. This was a full-blown forgery—and moreover a forgery that had created one of the most famous fossils in the world, the beautifully preserved Jurassic early bird, *Archaeopteryx*.

The fraud allegation against this poster-child of evolution, half reptile, half bird, had already made the news. I knew that the Natural History Museum in London, which houses the "type" or reference specimen, would now reluctantly be re-examining it in order to provide rebuttal. This task had fallen to Dr. Alan Charig, the Museum's cheery and gregarious curator of fossil reptiles and birds, who, with his curatorial colleagues through the ages, stood accused of concealing and enhancing the "forgery." He had already done the work in 1985—and, as Alan's collaborator Dr. Angela Milner wrote years later in Alan's obituary, was now "fulminating . . . at the enormous encroachment on his research time." At the time that I was asked to write my commentary, this careful study was about to be published after itself grinding through the process of peer review. I saw little point in reiterating the scientific conformation of what was, in any case, a foregone conclusion, so *New Scientist* agreed with my idea for taking another, non-scientific angle on the story— namely, the history of the *Archaeopteryx* specimens.

Creationists have long attempted to discredit *Archaeopteryx*, because it is exactly the sort of fossil that they like to pretend does not exist—one that shows evolution happening. Paleontologists hate the term, but *Archaeopteryx* is a missing link that is not missing. Hoyle and Wickramasinghe's accusations of forgery implicated Dr. Karl Häberlein, a Bavarian physician, who was the fossil's original owner. They suggested that Häberlein, to bolster Darwin's theory, had taken the remains of the small dinosaur *Compsognathus* from the Solnhofen Limestone, gouged out the rock around its bones, filled the hollow with some sort of soft plaster and, using a modern bird quill, imprinted feathers all around the skeleton.

Hoyle and Wickramasinghe's forgery allegation was itself nothing new. Similar scurrilous suggestions had done the rounds in 1862, when the large sum paid for the specimen by the Museum's controller, eminent anatomist Sir Richard Owen, raised eyebrows and led to much unwelcome comment. To allege that Owen wanted the specimen through a desire to bolster Darwin's theory would be absurd, since Owen hated evolution with a passion. Instead, Hoyle and Wickramasinghe alleged in their 1986 book on the subject that Owen had intended to expose the forgery later, in order to discredit evolution. However, and alas for Owen's dastardly plot, because Darwin and Huxley never commented on the fossil, it came to naught. As in the case of the Orgueil meteorite, the intended scandal never broke.

This whole absurd and convoluted conspiracy thesis relied upon so much special pleading of this sort that one is left wondering how two scientists, so fond of wielding Occam's razor themselves when attacking the things they oppose, could have been so blind to the overcomplexity of their own ideas. It is a testament to the power of too much belief, and that scientists are not immune to it—especially when they find themselves wandering without a guide in foreign territory.

Hoyle and Wickramasinghe's "evidence" was in fact so preposterous that it could have been refuted by any competent geology undergraduate able to tell plaster of Paris from lithographic limestone. Such a student would have had training and experience in distinguishing rock types that great theoretical physicists could be excused for lacking. Nevertheless, writing in the *British Journal of Photography* Hoyle and Wickramasinghe said that low-angle lighting and slow, fine-grained film used in their re-examination had revealed the feathers to be double-struck, as though inexpertly printed. Accusing the paleontologists of precisely their own failing, they concluded that the specialists were blinded by belief.

The real reason for their so-called double-strike effect was that *Archaeopteryx* bore its feathers in two slightly overlapping rows. All of the alleged evidence of forgery was, in due course, comprehensively debunked by the NHM team in an unequivocally titled paper in the journal *Science*, "*Archaeopteryx* Is Not a Forgery." The evidence of the fossils' history was, as I found out, no less convincing. London may house the so-called type specimen, but there isn't just one, single *Archaeopteryx* fossil in the world. These fossils, though rare, had been coming out of the Jurassic limestone of Bavaria since the 1820s, the first in 1855—anticipating Darwin's theory by four years. More specimens came to light in the 1860s and '70s, and the 1950s. So for their theory to hold historical water, Hoyle and Wickramasinghe's fakes must actually have been the product of a cottage industry, handing its skills down, perhaps father to son, for well over a century.

But why were two such eminent scientists led into making such a public display of themselves? And what, we might ask, had *Archaeopteryx* ever done to them? The answer lies in deep space.

Fred Hoyle had long been struck by the odd way many new viruses seemed to pop up in different parts of the world at the same time. For his part, Wickramasinghe held the belief that single-celled organisms could live in space and be distributed by meteorites or comets. The two men combined their theories in the suggestion that epidemics of influenza, for example, were visited on the Earth as it passed through trails of debris left behind by comets. Subsequently, in 1978, the two extended this epidemiological theory to suggest that throughout Earth history, deadly bugs from space had caused massive pandemics. The alien DNA in these malevolent microbes had infiltrated itself into the genomes of all earthlings and caused genetic storms that killed off many species and replaced them with new forms.

Here was a new explanation of mass extinction events. In advance of the Alvarezes' discoveries, they suggested that the K-T Boundary extinction had been one such event, which had not only allowed mammals but also birds to take dominion over the Earth. The reason that they chose to date the origin of birds to the K-T mass extinction and recovery is not clear. However, the fact that birds were already around in the Jurassic 150 million years ago was a tad awkward. Hence the need to discredit the fossil.

Rarely in the annals of science has the arrival of visitors from disciplinary outer space caused such controversy; and not since H. G. Wells's *War of the Worlds* have those alien visitors flopped so spectacularly for failing to do their basic research. (For those who have not read the novel or seen the adaptations, Earth is saved from Wells's aliens because the hugely advanced incomers had somehow overlooked that the Earth was crawling with infectious bacteria and viruses to which they had no immunity.)

Reading Hoyle's and Wickramasinghe's writings leaves any biologist or Earth scientist with a slightly queasy sensation. One comes across many lofty mathematical pronouncements, for example, about how evolution by natural selection cannot have been responsible for life on Earth because probability theory forbids it. One is reminded of Grove Gilbert and Barringer Crater, and of how the careful application of advanced mathematics to a set of mistaken assumptions led him astray—for the same forces are at work here too. Hoyle was evidently unaware that because it achieves its work by infinitesimal degrees, natural selection is not required to produce eyes and brains and other complex things all at once in a single bound. Hoyle likened the probability that evolution by natural selection could produce the results with which it was credited by biologists to the likelihood that a hurricane blowing through a junkyard should accidentally assemble a Boeing 747. In fact this whole logical nonsense, which ignores the incremental nature of evolution, has been dubbed Hoyle's Fallacy in recognition of its distinguished originator.

Hoyle's and Wickramasinghe's invasion of biology, paleontology and geology failed because the pair never properly familiarized themselves with the terrain they were invading. Their most frequently cited references on matters of biology and Earth science are Sir David Attenborough's *Life on Earth*, a BBC series tie-in, and Herbert Wendt's *Before the Deluge*, a popular book about the history of geology written in the 1960s, whose English translation appeared as a Paladin paperback in the 1970s. Both are excellent but hardly adequate groundwork for such an audacious and ambitious campaign. The only people to pay any attention to Hoyle's and Wickramasinghe's ideas today are creationists of one kind or another. Fred Hoyle died in 2001. Wickramasinghe still holds to panspermia, doubts the theory of evolution by natural selection, and recently appeared in a U.S. court as a witness on behalf of those who were demanding equal time for creationist theories in the classroom.

I once had the good fortune to sit next to Sir David Attenborough at dinner. At one point during our conversation I asked him how he felt about the prominent role he seems to have played in Professor Sir Fred Hoyle's biological education. After raising his eyebrows and inhaling deeply, he replied with a smile: "I don't know which is the more flattering: knowing that the great Fred Hoyle read my book, or realizing how little he understood it!"

✳

On August 7, 1996, at about 1:15 p.m., President Bill Clinton walked out onto the South Lawn of the White House with science and technology adviser Jack Gibbons, and for the benefit of the assembled TV crews commented on a paper in the journal *Science*. Presidential comments on scientific papers are perhaps as rare as *New York Times* editorials about uniformitarianism, and people took notice, as well they might. NASA scientist David McKay and his colleagues had just announced, at an earlier press conference, that they had (maybe) found chemical and physical evidence for life on Mars—or, to be precise, on an ancient meteorite fragment that had come here from Mars.

Meteorite AH84001's name and number that tells us it was the first meteorite to be discovered at Allan Hills in Antarctica in 1984. It weighed when found a little short of 4 pounds, and was originally misidentified as a rare kind of meteorite called a diogenite. The error was not corrected until 1993, when David Mittlefehldt of Lockheed Engineering and Sciences, Houston, measured the meteorite's oxygen isotope composition and found that it matched a very select meteorite clan indeed, named for the Shergottite, Nakhlite, and Chassigny meteorites (abbreviated to SNC and pronounced "snick" by aficionados). What makes these rocks so intriguing and rare is that they have been bounced to Earth from Mars, following impacts on the Red Planet, a fact proved by their having the same isotopic signatures as Mars's atmosphere.

Using various radiometric techniques, scientists established that AH84001 was formed by the solidification of previously molten material deep below the surface of Mars 4.5 billion years ago, when the Solar System was young. Subsequently it was shattered by impact, about four billion years ago, having worked its way closer to the surface by geological processes. There it remained, witnessing the evolution of Mars from a geologically active, wet planet, to the unremittingly cold and almost inactive desert world of today. Then, about 17 million or so years ago, it was blasted off the planet, from a site near Eos Chasma—a branch of the stupendous canyon system Valles Marineris, whose 2.5-mile-tall canyon walls expose a record of most of Mars's geological history.

Scientists have suspected Eos Chasma as AH84001's home since rocks with very similar reflection spectra were identified there by Vicky Hamilton of the University of Hawaii at Manoa in 2005. Moreover, close by lies a 12.5-mile-diameter impact crater—which indicates a bang big enough to have ejected the rock out of Mars's gravitational grip. Propelled into an Earth-crossing orbit, AH84001 wandered in near space until intercepting our planet about 13,000 years ago and falling on Antarctica's Far Western Icefield. There it remained, moving slowly with the ice sheets until, exposed by ablation, it was recovered by meteorite hunters from the ANSMET Project two days after Christmas 1984.

On August 6 and 7, 1996, meteorite AH84001 shot across the public firmament. Presidents do not step forth onto the dew-dashed lawn for any old chunk of Mars; but evidence of wetness, combined with possible fossils, was in media terms a dream ticket for President Clinton. Finding life on another planet would be the most significant scientific discovery of all time. To find it on Mars would open up all kinds of literally exotic ideas. For example, perhaps, during the stage in Solar System evolution known as the Late Heavy Bombardment, having originated and become established on some less battered haven like Mars, microscopic creatures may have then seeded the Earth with life—making us all part of a great Martian diaspora. For this sort of thing, presidents make room in their schedules; because for this sort of science the public will happily devote its tax dollars. And NASA, already with eyes on Mars, had more than a few ideas about how to spend them.

What McKay and his colleagues had found were small, rod-like objects, between 20 and 100 nanometers (billionths of a meter) long. This was smaller than any generally accepted living cell known on Earth—though the lower limit is going down as new forms are discovered, and some biologists claim that nanobacteria exist as small as 50 nanometers. The objects seemed to be built of shaped grains of magnetite—suggesting that a biological process might have been at work, similar to that seen in living magnetotactic bacteria, which make their own magnetite grains to help direct their movement. They also found globules of carbonate minerals, which looked tantalizingly like very small cells, and traces of polycyclic aromatic hydrocarbons—in other words, organic molecules whose carbon atoms are arranged in many linked rings.

For a few exciting years the idea that life might suffuse the cosmos enjoyed a galvanic twitch. Unfortunately, despite the boost it gave to public enthusiasm for NASA missions to Mars, skeptical voices raised doubts. The carbonate globules and magnetite rods, which many found so suggestive, could be the result of much less sexy inorganic mechanisms. As for AH84001's carbon-rich content, this could have come about as a result of mixing truly extraterrestrial organic molecules and good old genuinely organic terrestrial contaminants during the 13,000 years that have passed since it fell on Antarctica. Life may or may not be everywhere in the cosmos, but as investigations in the years since Bill Clinton's announcement have shown, it is everywhere on Earth—even on Antarctica's Far Western Icefield.

After enduring thirteen years of cold-water treatment, the organic origin of AH84001's mysterious blobs renewed its claim to acceptance in a paper published in November 2009. Detailed analysis of the magnetite grains had showed the inorganic alternative origin to be untenable, leaving the biological explanation—however improbable, as Sherlock Holmes might have said—still looking for effective disproof. The robust paper was written by a team led by Dr. David McKay of the Johnson Space Center, Houston, who had been part of the original investigating team that had put forward the biogenic hypothesis. One by one he dismissed skeptics' claims; for example, that the carbonates had formed at high temperatures, or that the organic molecules were Earthly contaminants. The minute magnetite grains had attracted particular attention because they were strongly reminiscent in form and composition to similar grains produced by some terrestrial bacteria that use them as tiny compasses to guide their movement. Skeptics maintained that these might instead be the products of intense heating affecting iron-rich carbonates, heating that might have been caused by the very impact that ejected the meteorite from Mars.

However these proposed inorganic mechanisms had not been supported by subsequent experiment, the team concluded. In fact, the observed chemical and crystallographic properties of the grains would have been enough to label them as biogenic, had they been found on Earth. Another objection, that Mars today does not have a strong magnetic field, rendering magnetite grains useless to any bacterium, has also evaporated. Satellite mapping has shown that many ancient Martian rocks of the same age as AH84001 possess strong magnetization. This could only happen if, early in the planet's history when the supposed bugs would have been alive, a planetary magnetic field like that of the Earth did indeed exist. McKay and colleagues remain convinced that no other explanation of the evidence as presently understood is as persuasive as the organic explanation.

Evidence for fossil Martian life-forms in AH84001 seems to have come through its first tests. However, though the possibility remains open, given that the evidence we have is being asked to support what has been described as the most important scientific revelation of all time, it is simply not extraordinary enough—yet. Bugs from space is still an idea whose time, if it ever has one, is not now. We can file it as yet another thing we do not have to worry about.

But this is not the end of the idea that, under the right circumstances, meteorites can stimulate biodiversity. This indeed is the surprising conclusion of a series of discoveries that began with Professor Per Thorslund and a mysterious slab of rock in the basement of his university department.

8

RAIN FROM HEAVEN

The earth shall bear stars.

JOHANN BERINGER (1726)

In 1952, Swedish geologist Per Thorslund of the University of Uppsala received an unusual present: a polished slab of limestone about 25.5 square inches in size. It had been found at the Gusta Stone Factory in Brunflo, where workers had been sawing stone slabs from the Rödbrottet quarry not far away, near the hamlet of Gärde in central Sweden. Stone like this has been used in Swedish buildings since the twelfth century, so you might say its characteristics were well known. This particular slab had caught the cutter's experienced eye because, as well as fossils common to Ordovician-age limestones of that area, the brownish-red rock, which had given the quarry its name, contained a black object about 4 inches across, extending through its entire 1.2-inch thickness and probably beyond into the neighboring slab.

Using the same basic techniques developed a century before by Henry Sorby in Sheffield, thin sections of the mysterious object three hundredths of a millimeter thick were prepared on glass slides and examined under a petrological microscope. The petrographer reported that the unknown black object was a "highly metamorphosed ultramafic rock." What these words convey to a geologist (and conceal from anyone else) is that it was made up of the dark minerals that are typical of rocks that form in the mantle of the Earth under conditions of intense heat and pressure.

Such rocks do occur in parts of Norway and Sweden because they form part of the ancient basement of a long-lived and extremely stable continental unit called the Fennoscandian Shield. The Earth's continents, as we see them today, all possess ancient cores, formed thousands of millions of years ago. These shields have been drifting about the surface of the Earth for billions of years, colliding with one another, annealing new material to their outer edges, and breaking apart again as ancient supercontinents fragmented and re-formed. Ultramafic rocks are remnants of the ocean crust and underlying mantle that get caught up in this process.

Ordovician rocks, which blanket the ancient shield rocks in areas of Sweden, formed in a shallow sea that then covered the area, rocks that were subsequently protected from Earth movements, which take place mainly at shield edges where collisions occur, trapping younger sediments in the jaws of a vice. British geologists, accustomed to the mangled Ordovician of their own country, are often quite shocked when they first see rocks of the same age in the Baltic regions—I certainly was when I began my own Ph.D research there on younger, Silurian strata. The rocks of this age that they are used to seeing consist of dark, deep-water sediments which became caught in the mangle as the Fennoscandian Shield collided with Canada and North America—and so built the mountains whose eroded roots can still be seen in the hills of Wales and Scotland.

Contrastingly, Sweden's relatively shallow-water rocks—limestones for the most part—formed in temperate, tranquil seas no more than 328 yards deep. Here they accumulated very slowly as shelly organisms died and settled, together with a little wind-blown dust, on the sediment-starved sea-bed. Although hard (limestones lithify very quickly), these rocks seem to belie their half-billion-year age, looking almost as fresh as the day they were deposited. Still lying almost horizontally, cross-cut by very few fault lines, single beds—mere centimeters thick—can often be traced across thousands of square miles.

Because of the slow deposition rate, geologists refer to these rocks as a condensed deposit in which small thicknesses of sediment represent a lot of time. It is as though the tape-recorder of accumulating material ran very slowly for millions of years, taking, on average, a thousand years to lay down just .08 inch of limestone. For this reason, finding an embedded 4-inch-diameter boulder—of no matter what composition—many miles from the ancient shoreline was unexpected to say the least.

In certain circumstances you can find large boulders sitting in otherwise fine-grained sediment. Around the edges of ice sheets, for example, melting glaciers and icebergs drop cobbles and boulders into deep water, becoming what geologists call dropstones. Often the laminae of the fine sediment in which they landed can be seen to have been bent downward, as the plummeting boulder punched its way into the mud of the seabed. But glacial dropstones are rarely isolated. Where one is found, there tend to be others. Also, the prevailing climate when these limestones were deposited was typical of temperate mid-latitudes. The context seemed to rule out a glacial origin. In the end, after much racking of brains, Thorslund concluded that nevertheless the mysterious object simply had to be *some* kind of dropstone. Perhaps it had been rafted offshore on a floating mass of seaweed?

But this was also unsatisfactory. Thorslund knew that the location where the stone had somehow fallen to the seabed had been located in an embayment in the eastern shore of a long-vanished ocean known as Iapetus. He knew that the water there had been quite shallow. He also knew that along that vanished 464-million-year-old shoreline, there were no outcrops of highly metamorphosed ultramafic rocks that might have acted as its source. And so, as with many such insoluble conundrums, because nobody could think where to take the investigation next, the perplexing slab was set aside—for twenty-five years.

But it was not forgotten. As I saw for myself on one memorable field excursion, scientists reserve a particular love-hate for those things they cannot even begin to explain adequately. A common feature of certain limestones is a streaky veining effect called *Stromatactis*. It had long been given its Latin name, in the belief that it was a fossil; but was it really? Or was it a feature created after deposition? Or was it maybe a bit of both? When I was a student there were many theories about this, and each had its points; but nobody knew. And so I remember, during a trip to the Pennines of northern England, looking on as the great limestone expert of the time, Dr. Robin Bathurst of Liverpool University, tore off his jacket on a freezing morning, just to conceal a particularly spectacular outcrop of the stuff, shouting: "Oh, cover it up, cover it up—it's so *embarrassing*." This was rather how Per Thorslund must have felt about the mysterious slab from Jämtland.

Often what is needed to break an impasse like this is a conceptual leap—and this came in the 1970s, as geologists began to take heed of Derek Ager and others, as they rehabilitated the catastrophe in geological history. Lying about 124 miles to the south of Brunflo is a circular geological structure called the Siljan Ring, which was only then being recognized for what it was. The news must have streaked through Thorslund's mind; it certainly awakened his interest in meteorites, and it did so just as he was rearranging his working collection in December 1979. He found himself being struck by a wild surmise. Perhaps that mysterious black dropstone had indeed dropped, but from a very much greater height.

Thorslund re-examined the object, as he wrote, "with the idea that it might be a meteorite." And sure enough, the simple fact of believing it possible rendered many things newly visible. Thorslund, and colleague Frans Wickman from the University of Stockholm, noticed that the mass had a clotted nature; they realized that they were looking at the outline remains of relict chondrules. Closer inspection revealed many characteristic chondrule structures—one showing radiating crystals, another parallel bars, and many more with distinctive outer rims.

Thorslund's tantalizing object turned out to be the first definitive example of a meteorite fossilized in earthly sediment—a shooting star, caught in stone. Once recognized as such it was given the name of Brunflo, from the nearest town to the quarry where it had been unearthed a quarter-century before. But Thorslund and Wickman were not, they knew, looking at original meteoritic minerals. Brunflo is a true fossil, whose original substance has been dissolved and replaced, molecule by molecule, preserving original structures in new material. This is exactly comparable to what happens when dinosaur skeletons, for example, become petrified and change from bony, phosphatic originals into the durable mineral calcite. Brunflo's original meteoritic minerals, such as olivine, pyroxene, nickel-iron, and a characteristically meteoritic form of iron sulfide called troilite, had vanished. In their place, Thorslund and Wickman found calcite, barite, phengite (rich in chromium and vanadium), and the mineral cobaltite. Only one primary, extraterrestrial mineral appeared to remain more or less intact—a supremely resistant substance called chromite.

Thorslund was clearly excited by his discovery—past even the usual joy at having resolved a nagging issue. A fossil meteorite was a spectacular find, and, getting just a little carried away, Thorslund noted the presence next to it of a fossil nautiloid (a distant ancestor of today's *Nautilus,* whose shell was a straight cone, instead of being rolled up) and suggested that perhaps it had been struck and killed. Alas, this exciting scenario is not very likely. Fossil nautiloids are common in these limestones, and the seabed on which the meteorite settled, and it would probably have settled with little real force, would have been strewn with accumulated nautiloid remains, so it is not surprising that Brunflo simply landed next to one of these shells or possibly that the shell landed next to the meteorite a few thousand years later—there's no way of knowing. However it and other, similar shells, do confirm that seabed conditions must have been very calm indeed. The slightest current activity would have ensured that such long, light objects were preserved pointing in the same direction—that of the prevailing current. No such preferred orientation has ever been found. The Brunflo meteorite lay exactly where it fell although, during the intervening 464 million years, the tectonic plate beneath it had carried its passenger thousands of miles, across the equator, almost from one pole to the other.

For Thorslund, the discovery marked the crowning achievement of a long career. The announcement was published in

January 1981 in *Nature*. However, the promised follow-up paper, giving a detailed description of the world's first definitive fossil meteorite, did not appear until 1984, and was completed by other hands. Professor Per Thorslund died, after a brief illness, only eleven months later.

In Lewis Carroll's poem "The Walrus and the Carpenter," the mischievous pair experience great difficulty at first in persuading the oysters, on whom they plan to feed, to accompany them on their walk along the briny beach. But once one young oyster hurries up, eager for the promised treat, it is not long before the two have as many as they can eat. Something similar soon happened in Sweden. Brunflo might have remained an isolated curiosity had not yet another fossil meteorite turned up in 1988, in slightly older limestones (467 million years) at the Thorsberg Quarry near Kinnekulle, 373 miles to the south of Brunflo. But this is to get ahead of our story.

Between 1970 and 1979, just about the time that Thorslund was re-evaluating his slab, I was beginning my Ph.D research in Sweden and briefly met the professor during a whistle-stop tour of Stockholm and Uppsala. And, just as I was finishing my project, a Swedish research student named Birger Schmitz embarked upon his own Ph.D project at the University of Lund. His research concerned the chemistry of the oceans and how this was expressed in their deposits; but to the surprise of his conservative academic peers, Schmitz read the newly published paper on the K-T Boundary by Luis and Walter Alvarez, Michel, and Asaro—and decided to take it seriously. In fact, he took it seriously enough to decide that he must change his whole research topic.

In the hidebound, clubby, and distinctly old-fashioned geological establishment of Sweden at that time, one can imagine the reaction. "They were—not very positive, shall we say," Schmitz told me. "They said it was all *geofantasy!*" But Schmitz had made up his mind. Some of the best K-T Boundary sections lay practically on his doorstep. With the instinct of a true scientist—an instinct not that different in principle from what in a journalist is called a nose for news—Schmitz realized it would be crazy not to use them to test the biggest scientific idea in town, even if it did prove to be geofantasy.

His peers relented—as indeed they must if a research student is really determined—and, while writing his newly titled thesis, Schmitz committed yet another daring act destined not to please them. He published two sole-author papers about the Alvarez hypothesis: one in the prestigious journal *Geochimica et Cosmochimica Acta,* another in the highly regarded *Geology*—as well as another in *Cosmochimica* in which he was senior author. The first of these papers, published in 1985, was particularly critical of the Alvarez hypothesis; for, based on his understanding of how chemicals precipitate from solution, Schmitz could not reconcile the existence of the iridium anomaly with the idea of its sudden extraterrestrial introduction.

By coincidence, about this time Luis Alvarez came to Stockholm, where he was invited to dinner by the Swedish Academy. Luis surprised the Academy by asking if they could ensure that this young Schmitz character could be invited along. Not many research students publish sole-author papers in *Geochimica;* even fewer get invited to dinner at the Swedish Academy at the behest of a Nobel Prize–winner. Of course, the inevitable happened. "They put me really far away, right at the end of the table," laughs Schmitz. "Luis was surrounded by all the big professors that the Academy had invited, but the person he really wanted to speak to was me!" Eventually, at coffee

afterward, they did manage to talk. The encounter, while combative, evidently won him Alvarez's respect, with the result
that Schmitz was invited to do his post-doctoral research on
the K-T Boundary at Berkeley.

The evidence of shocked quartz grains—grains of silica that
have had their crystal lattice distorted by intense pressures—
was winning most people over to the Alvarez hypothesis by
the time Schmitz moved to Berkeley in 1988. Shocked quartz is
taken as an unequivocal sign of impact because no terrestrial
process, even explosive eruption, is thought capable of inflict-
ing such damage. Schmitz came to realize that, while he had
been right about the geochemical details, he may have been
wrong about the bigger picture. He became a believer, and
returned to Sweden mentally prepared for what was about to
emerge, thick and fast, from the Ordovician sea.

Schmitz soon heard about the discovery of the second
Ordovician fossil meteorite at Thorsberg. Although the rocks
were slightly older, they were very similar—reddish, condensed Ordovician limestones, rich in fossil orthocones. A
local geologist, Mario Tassinari, came to hear about it through
his local newspaper in Lidköping and was quickly on the scene
asking the quarrymen if they knew of any more. Before long,
Schmitz was also on hand and together with Tassinari began a
program of systematic sampling.

Since that sampling project started in 1993, ninety individual meteorites have been recovered. "It is remarkably pre-
dictable," Schmitz told me, in 2009. "Each year in Thorsberg
Quarry, another 820 square yards of Ordovician sea floor is
quarried, and each year it has produced four to eight meteorites—six on average. Up till now, 13,123 square yards have been
excavated. Of the 90 meteorites found, 65 have been analyzed

and all have been diagnosed as L chondrites." Schmitz admits that this side of his work is becoming a little boring. "I think there is a very high probability that they are all L chondrites!" he says.

L chondrite meteorites are among the most common meteorites among present-day falls. They are undifferentiated meteorites with a distinctive low-iron chemistry whose chemical, isotopic, and textural signatures make it certain that they all derive from the same parent body. Thorslund's Brunflo meteorite, after being correctly identified as a meteorite, had been wrongly designated as an H chondrite (another common meteorite group). However, re-assessment by Schmitz and colleagues has since established that it, too, was an L chondrite.

For Schmitz, the scientific turning point—the moment when his investigation changed from a geological curiosity to a major interdisciplinary project with implications for Solar System history, meteoritics, astronomy, and even evolutionary biology—came after they had found about nine meteorites at Thorsberg. "Suddenly I thought—hell, isn't this a *lot* of meteorites?" he recalls.

The question of what constitutes a lot of meteorites centers on something called the Earth's meteorite flux—basically, the normal background rate at which material arrives here from the cosmos. At the time, the young meteorite scientist Phil Bland, now with Imperial College but then at the Natural History Museum in London, was publishing results from his work on the Nullarbor Plain of southern Australia. This is an ancient desert area, where finding fallen meteorites is relatively easy because the background surface is very light in color. Bland's work at that time involved dating the surfaces underneath the meteorites he discovered, to derive some idea of the fall rate.

"I did some quick calculations comparing areas, sedimentation rates, numbers of meteorite finds, and so on, and I quickly decided that the flux must have been amazingly high at this period in the mid-Ordovician," Schmitz says. Subsequent work has confirmed that in the three million years after the start of the bombardment (468 million years ago), the flux rose several hundred times, and subsequently remained ten times greater than it is today for a further ten million years after that. Knowing the area covered by their searches, estimating the original weight of meteoritic material that it represented, and relating that to the time intervals represented by the rocks, Schmitz and his colleagues came to the startling conclusion that at its peak, meteorite flux reached *at least* 100 to 150 times its present value.

By no means did all of Schmitz's finds consist of actual hand specimen meteorites. Much of the most significant evidence came from single grains of that resistant mineral chromite, the only true remnants of original meteorite material still extant in the fossil meteorites studied in thin section. Using acid baths, these grains were carefully extracted from limestone samples collected over 155,343 square miles of Sweden. Because they had survived the ravages of time relatively unaltered, Schmitz and his colleagues were able to perform detailed chemical, and later isotopic, analyses on them. The grains all showed the same chemical composition—not only as each other, but also as the chromite grains found within the meteorites. Also, these tiny, sand-sized particles were not, apparently, the remnants of once-large meteorites that had been destroyed on the sea floor, or in the Earth's atmosphere. They had arrived as micrometeorites. The Ordovician Earth had been sprinkled with stardust.

Using these tiny relics of meteoritic material, Schmitz and co-workers soon discovered that the chromites' cosmic ray exposure age increased up the section, from older to younger rocks. Exposure to cosmic rays in space implants certain unique isotopes (varieties of a chemical element which have different numbers of neutrons in their nuclei, and so possess different masses) into the superficial thicknesses of all meteorites. Individual chromite grains which Schmitz sieved out of the acid-bath residues also showed enrichment in these distinctive isotopes. From this it followed that even the tiny grains must have been floating around in space—and not shielded within bigger objects. Yet as more and more cosmic ray exposure age results came in, it became clear that the younger the terrestrial age (i.e., the later they fell to Earth), the longer the meteorites and micrometeorites had spent in space. This meant that all the grains and meteorites had originated in one catastrophic event—which had liberated them all and simultaneously begun their exposure to those cosmic rays. That event was the destruction of the parent body of the L chondrite meteorite clan.

The fact that closely clustered shock ages of L chondrite meteorites might be explained by a single large parent-body break-up event at that time, was first proposed by meteorite scientist Edward Anders in a major paper published in 1964. If true, this would have resulted in the Earth's being bombarded, for several million years, by meteorites ranging in size from the very minute to perhaps the enormous. The time period covered by the raised meteorite flux also encompasses the estimated ages of no less than four major meteorite craters in Baltoscandia alone—at Lockne, Kärdla, Tvären, and Granby.

If this theory was correct, then the effects of the disruption should have been felt all over the globe, and Schmitz and his co-workers lost little time in trying to prove it. The first con-firmation came from not very far away—in southern Sweden; but later, coeval rocks from central China, exposed along the banks of the Puxi River, showed similar increases in microme-teorite content.

Nor can we escape its consequences today. Twenty percent of all modern meteorite falls are of the shocked L chondrites that originated in that same break-up event. We are still wit-nessing the long tail of what Schmitz has called "the biggest bang in the Solar System for a billion years." The event even touched that young soccer player in Mbale, Uganda, on August 14, 1992, when a half-inch-sized fragment from a much bigger meteorite that had exploded high in the atmosphere landed on his head. That object's final plunge had been set in motion back when no living thing more evolved than orthocone nau-tiloids plied the waters of Iapetus.

Just to account for the amounts of meteoritic material they had found, Schmitz and his co-workers estimated the size of the L chondrite parent to have been at least 62–93 miles in diameter. The destruction of such a large body would also have resulted in the creation of a large asteroid "family" orbit-ing together somewhere in the Asteroid Belt. The next ques-tion was, which family? Originally, attention centered on the Flora Family, which is situated in the inner main Belt.

We have already seen how meteorites in Earth-crossing orbits reach our planet once they become expelled from the Asteroid Belt by straying into one of a number of clear zones—the Kirkwood Gaps. The Flora asteroid family lies close to one such gap; but as a candidate for being the remnant of the L chon-drite parent-body break-up, it had some problems. The most significant was that the Flora family's reflectance spectrum, by which astronomers are able to determine the minerals present on the surfaces of asteroids, did not match the mineralogy of L chondrites.

As recently as 2009, however, another flock of related asteroids, the Gefion Family, was found to provide a much better fit. The age of the family had already been estimated from other evidence at 485 million years (plus 40 or minus 10 million years), which was in the right region. Moreover, the Gefion spectrum suggested a better fit with L chondrite mineralogy. Also, the diameter of the family's parent body—determined by putting back together all the asteroids in the family today—turned out to be between 62 and 93 miles. As asteroid families inevitably become winnowed down, this was a minimum figure, and once again it fell in the same broad region as the estimate developed by Schmitz and colleagues based on meteorite evidence.

The cosmic ray exposure ages observed in the fossil meteorites were, at their minimum, extremely short. This meant that they had not spent very much time in space before they first arrived on Earth—something between a few thousand or tens of thousands of years, and a million or more years. So, if the Gefion Family was indeed the source of the L chondrites, then to make such low cosmic ray exposure ages possible, its members would need to be fairly easily diverted into resonant orbits (i.e., propelled into the nearby Kirkwood Gap). Once there, they would also need to possess orbital properties necessary to make a short transit time to Earth possible.

Computer simulations have now shown that both criteria are met for Gefion fragments. A Gefion object needs to achieve escape velocities of only 55 yards per second, which is nothing in cosmic terms, in order to achieve a resonant orbit, even if starting from deep in the heart of the Family. Once in the Gap, simulations show that a fragment could reach Earth in as little as 50,000 years. And in a final confirmation, after the initial collision that created the Gefion Family, the rush hour period, when the flux of incoming fragments falling to Earth would be expected to peak, would be about one or two million years—which is exactly what Schmitz and his co-workers have determined independently from the volumes of meteoritic material found in the mid-Ordovician rocks.

A truly good scientific hypothesis does two things—it ties together many observations, often from widely separated scientific disciplines—like Edward Anders's explanation for the shock ages of L chondrites—and resolves long-standing mysteries, such as Per Thorslund's perplexing dropstone from Brunflo. But it also does something else. Just as the Alvarezes' hypothesis about the K-T Boundary did, but which the idea of dinosaurs dying of constipation or overheated testicles did not, a truly fruitful idea generates tests for itself.

Such ideas point toward new avenues of research. With the help of a little discipline-crossing publicity, scientists in even superficially unrelated areas realize there is something they might be able to do to help test the theory. They then return to their own data with a new idea in their heads. They reinterpret the old within a new conceptual framework; they look again, believing now in the possibility of something different. A new idea, coming in from way beyond the confines of their own little world, seems to light up the night sky and make the trees whisper.

Ideas that change your worldview may not, of course, be new to everyone at the moment you find out about them—they may just be new to you. Educators refer to such ideas as threshold concepts—ideas that change forever the way someone sees the world. For this reason they are also called troublesome knowledge—because anything that causes you to re-assess radically your place in the universe inevitably is. The defining characteristic of such ideas, as of any true revelation, is that they are Damascene—once grasped, they are irreversible. You cannot unknow that the Earth is round, or that it orbits the Sun, or that it is 4.567 billion years old, or that the universe is so much older still that other Solar Systems may have formed and been destroyed long before ours ever came into being. Like the Apostle Paul, you are changed forever.

My own exposure, back in that accidental impromptu chemistry lesson, to ideas of stellar nucleosynthesis and the meaning of the periodic table had been such a threshold for me. But when that same knowledge came newly minted, first in Fred Hoyle's mind and then elaborated in the great paper he wrote with the other members of the B^2FH quartet, it was something else. When a great idea—or sometimes, a great idea coupled with particular evidence, as in the case of the Alvarezes' hypothesis—is truly new to everyone, the revelation occurs en masse. Hitherto puzzling anomalies become new building blocks in the great Babel-tower of science. Pieces suddenly start to fit together where before their connection had seemed impossible. Scientists everywhere stop what they are doing. Research students announce that they are changing their topics, and hidebound colleagues find themselves appalled.

Philosophers of science call such conceptual revolutions heuristic, to recognize the way they tend to lead to new discovery. The word may be unfamiliar, but it comes from the same Greek root as Archimedes' famous cry, "*Eureka!*" or "I have found it!" Scientists as a whole—especially young ones, who are too busy doing science to worry much about how they

are doing it—do not on the whole know very much philoso-
phy, any more than these days they know Ancient Greek. But
a good scientist can tell a good theory when he or she sees
one; and that perhaps is enough. The better the scientist, the
quicker this instinct is to kick in, and the more determined
they are, like the young Birger Schmitz, to follow the new star.

Many problems remain to be resolved about the collision
that created the Gefion Family. For instance, it has probably not
escaped your attention that it takes two objects to make a collision. If one of these, presumably the bigger, was the L chondrite parent, what was the other, and where is it now? Schmitz
thinks this problem might be resolved by looking more closely
at inclusions in L chondrites—possible fragments of the other
impactor. But what most excites him is the prospect of dissolv-
ing even more of the stratigraphic column in baths of acid.

"We now have a tool," he says, referring to the extraterrestrial chromite grains that showered the Earth in heightened
numbers for perhaps 10 to 15 million years of the mid-Ordovician, "that can enable us to work out precise meteorite flux
rates throughout the geologic column. This is amazing!" But of
course, what this implies is a major project, taking samples from
sediments spanning much more of the history of the Earth, and
performing the same analysis on them as Schmitz et al. have
done on the limestones of Thorsberg or the Puxi River section
in China. It promises to be a major, almost industrial, undertak-
ing. Schmitz has plans to get going as soon as funding can be
found. "We're going to need a lot more acid," he says.

✳

Louis Pasteur, in a lecture of 1854, said famously: "In the fields
of observation, chance favors only the prepared mind"—which

is another way of saying that seeing is predicated upon belief. This was exactly what happened when Schmitz came back from Berkeley, mentally prepared for his later work on fossil meteorites. However, the ramifications of what he had uncovered would lead even further—scientifically speaking—than the discovery of a great collision in the Asteroid Belt 480 million years ago and the linking of the L chondrites, with their long-known shock ages, to the Gefion Family asteroids.

Many more mysteries of the unexplained were lying around, in other fields of study, waiting to receive conceptual fertilization from the wider universe and at last unite in a wider understanding of our Earth and life upon it as parts of the greater cosmic environment in which both exist. The simplistic assumption that cosmic influences cannot be anything other than disastrous for life, just as they appear to have been, 65 million years ago, for the dinosaurs, was about to be challenged. Meteorites need not—cannot—have just one, single meaning in the history of the Earth. Just as they have always derived historical meaning from the context in which they occur, science was on the verge of discovering that the evolutionary and environmental context into which meteorites arrive could be just as important in determining the meaning of those impacts for life on Earth.

As the hoax that ruined Dr. Beringer of Würzburg developed, the range of images the fraudsters carved upon the stones with which they salted the slopes of Mount Eibelstadt for him to find grew ever more fantastic. As well as representations resembling true fossils such as ammonites, Beringer also found bees sitting on the honeycomb, spiders lying in their webs, and frogs fossilized in flagrante. Their ingenious carver, whom one can imagine wondering just how far his victim's credulity could be pushed, turned his hand to inscriptions—in Latin, Arabic, and Hebrew—and eventually to depictions of meteors and comets. Still, Beringer believed that they were genuine.

Although old Beringer does not seem to have had much of a sense of humor, it might have given him some satisfaction to know that, in the second half of the twentieth century, scientists in Sweden would discover how wrong his many mockers had been to scoff at the idea of a fossilized shooting star. Perhaps, duped though he was into seeing the name of the Creator written in his lying stones, he would have recognized with satisfaction this overdue reconnection of our small world with the universe that gave it birth—and in which it will always have its being.

9

A KICK IN THE GENES

I have always held that life is, or should be, more interesting than death.

DEREK AGER

Around about the same time that Birger Schmitz was publishing youthful papers in *Geology* and *Geochimica et Cosmochimica Acta*, I was attempting something similar. I was writing up my research on rocks from the Silurian Period, as exposed on the sunny Baltic island of Gotland. Like the older Ordovician rocks of the mainland, which one day yielded their rich haul of fossil meteorites, these rocks also seemed hardly changed from the day when they were laid down. They were barely broken at all by faults, and they dipped by less than one degree to the southeast. In fact, the regional dip of Gotland's rock strata is a mere 30 minutes of arc—half a degree—a fact of which I was rather boastful before my contemporaries, many of whom had to wrestle with multiple phases of faulting, folding, and metamorphism under the grim skies of northern Norway.

The rocks I was studying, called the Visby Formation after Gotland's walled Hanseatic capital, were the oldest on the island, and consisted of thousands of thin alternating limestone and mudstone layers about an inch thick, laid down in a gradually shallowing sea 430 million years ago. By then 50 million years had passed since the mid-Ordovician meteorite shower, and I was studying the way that the Silurian equivalents of today's coral reefs had begun to grow from small beginnings in deeper water into some truly gigantic structures higher in the succession, as the Silurian sea had receded. Small reefs, which had begun growing in deeper, quiet water, had come into contact with progressively shallower and more energetic waters, finally near the top of the Formation undergoing a sudden and spectacular diversification—turning from small and not very species-diverse patch reefs into the massive, cliff-forming structures that my colleague Nigel Watts was studying in the unit above. Our two research projects were well balanced and played to our abilities—or rather to my lack of them. All I had to do, to see my miniature reefs, was stroll along the beach. Watts, a mountaineer, spent much of his day dangling at the end of abseil ropes. My main hazard consisted of his rocks falling on my head. His main hazard was falling off his rocks onto mine.

Four hundred and thirty million years ago, modern corals had not evolved, so these Silurian reefs were built mostly by two major groups of organisms. One—the tabulate corals—is now extinct, while the other was a group of rather mysterious beasts called stromatoporoids. In the late 1970s, these were still widely believed to have become extinct at the end of the Cretaceous, when they were thought to have vanished forever along with the dinosaurs. However, during that decade living examples had begun to turn up in deep water, and nowadays we speak of stromatoporoids in the present tense. Most biologists agree that they are a form of highly calcified sponge. Just as corals do today, in the Silurian and Devonian particularly they laid down skeletons that could vary bewilderingly in shape according to species and environment. Most of those in the Visby Formation were shaped like simple jelly molds or upturned cereal bowls, approximately half an inch tall.

The very earliest evidence that I discovered of the impetus toward a reef-forming habit (organisms growing on other organisms) was the encrustation and boring of these stromatoporoid skeletons by a host of other beasts. Consequently I spent a lot of time, back in the lab, cleaning and describing these encrusted skeletons that had once sat on the sea floor like mini-reefs. I looked at evidence for ecological succession—determining which organisms tended to colonize first, which last. By comparing the encrustation patterns of each species with those left by modern encrusting marine organisms (such as barnacles and oysters, whose larval behavior is well known) I also did what I could to bring back to life the larvae of these long-dead cemented organisms. From the moment that their larvae make that once-and-for-all decision to settle, encrusting organisms are committed for life to one location. A barnacle larva therefore knows, because evolution has programmed it to know, that location is everything. The trick to being a successful barnacle is not very different from that of being a successful newsagent—you need to be where the trade is.

While writing my thesis, I also composed what I considered a nice little paper about all this, and had the temerity to submit it for publication in a top-end American journal called *Paleobiology*. I am afraid I was not as successful in my act of youthful hubris as Birger Schmitz. After receiving a formal acknowledgment, I endured some months of uncertainty while my manuscript was sent out to two experts for peer review. Eventually, my lovingly prepared typescript, together with camera-ready plates of laboriously hand-produced black-and-white photographs and line drawings, came back with a polite rejection letter signed by the journal's new editor, Jack Sepkoski, a pioneer in a new field of paleontology who was the first to integrate information about fossil species into a vast computer database. His work was to reveal a hitherto unremarked peculiarity about life in the mid-Ordovician that made Birger Schmitz realize that his fossil meteorites, and the asteroid collision that brought them, had had a profound—and profoundly unexpected—effect upon life on Earth.

Having a paper rejected, especially at the beginning of one's research career, is no big deal (and if you think it is, you are in the wrong game). But in this case, there was for me a coded message hidden in the signal from Jack Sepkoski in Chicago. I had rather shut my eyes to it, I confess; but a revolution had overtaken not only my chosen sub-discipline of palaeoecology but all of paleontology, a revolution nowhere more in evidence than at a go-ahead journal like *Paleobiology*. One of the two referees' comments indicated how outmoded my old-fashioned, descriptive science had become. At the top of the paper, he had written: "Not bad—reminiscent of a good grad student project." Faint praise, maybe; but praise is rare enough in the competitive world of scientific research, and you take it where you find it. Any pleasure I derived from this was doomed to be short-lived. Next, my reviewer delivered a decisive kick to the vitals by concluding: "Would have been perfect in the "BS" era," adding beneath, with a knowing wink: "Not what you think, Jack! 'BS' means Before Sepkoski!"

Jack Sepkoski combined being brilliant with a phenomenal capacity for hard work, and he had set about a quantitative revolution that was to transform the subject of palaeoecology. In his hands palaeoecology changed from narrative studies of the small and specific, like mine, into a highly statistical and theoretical science, computerizing published data so as to enable ambitious high-level generalizations to be made about the history of life on Earth. My miniaturist, descriptive approach, full of close-up photos of specimens and nary an equation in sight, had become quaint—so much "Before Sepkoski." Although I was not quick enough to see and understand in a flash that the future no longer belonged to me or my kind, and although my little paper was eventually published elsewhere, the writing was on the wall.

The habitat in which I, as a larval scientist, was trying to settle, was changing. I had to get out before I became irrevocably cemented in an unfavorable environment; before the coming revolution would sweep away the old world in which I was comfortable. When species are challenged by environmental disturbance, they move on and diversify. I had to do the same, and find an ecological niche where things I could do well were more useful, and the things I was hopeless at less fatal. And so, after a short spell in the halfway house of the oil industry, I finally crossed the border on a moonless night, at a weak point in the middle of nowhere—and became a science writer, observing scientists instead. For helping me make that escape I owe Jack Sepkoski and one anonymous reviewer a great debt of gratitude.

While still a research student himself, Jack Sepkoski had begun a monumental task that occupied the rest of his life (cut tragically short in 1999 by a heart attack brought on by high blood pressure, aged only fifty). Mining the voluminous literature of paleontology, he created a series of computer databases detailing the appearance and disappearance of fossil organisms since the advent of abundant fossils about 580 million years ago.

Sepkoski's work—like that of his Chicago co-worker David Raup and his Harvard Ph.D supervisor, the evolutionary theorist and essayist Stephen Jay Gould—blew a gale of change through paleontology, and amazingly he had achieved this revolution by the time he was a mere 35. As his former student Arnold Miller wrote the year his mentor died, Sepkoski was a remarkable revolutionary. This was not because, like Derek Ager, he was a reluctant one, but because of how willing he was "to endure the resistance of colleagues who had a difficult time adjusting to the changing landscape of paleobiology." I remembered, when I first read that, the gentle and considerate letter that Jack had written to me. Miller goes on: "I witnessed first hand just how difficult it was for Jack to overcome the inevitable prejudices that come with doing something truly new. Others . . . would have folded, or . . . found a way to ignore the withering comments and bickering to which he was subjected. Jack could do neither, and he went out of his way, formally and informally, to engage his critics in meaningful dialogue aimed largely at improving his data and interpretation."

Sepkoski started compiling his databases as a sideline while doing doctoral research on Cambrian fossils at Harvard under Gould. Delving into the source literature of the science of fossils, Sepkoski set about documenting the ranges in time over which groups of organisms existed, from their first to their last occurrence. By any standards this was a heroic task of data inputting, but was by no means as routine and automatic as that description would imply. In dealing with this mass of data, some of which is old, much of which might be incompatible, Sepkoski had to make expert judgments about what to count, and how to count it. Faced also with the fast-changing world of computing at that time, it was a minor miracle that he eventually published his *Compendium of Fossil Marine Families* in 1982.

The family, at least in the sense of a classificatory grouping in biology, enjoys very little public profile. The basic unit in all biology is the individual. A population of similar individuals who can interbreed to produce viable offspring is called a species, the fundamental unit in all biological naming and classification taxonomy. Because the species relies for its definition on the quintessentially biological activity of reproduction, it is slightly harder to define for fossils, which are about as past that sort of thing as it is possible to get. However, as every biologist will admit, things that look the same (that is, share certain key physical features) may be supposed to be the same. As the saying goes, if it looks like a duck, walks like a duck, and quacks like a duck, it is probably—in England, at any rate—*Anas platyrhynchos*. And despite the fact that frustrated male ducks will occasionally drown farmyard chickens by trying to copulate with them in the water, birds of a feather generally do what nature intended to their own kind. It is therefore a reasonable inference that if similar-looking creatures, now fossils, had once lived at the same place and time, those individuals could have bred, and therefore were con-specific.

Similar species are then grouped into genera—our own species (*sapiens*) belonging to the genus *Homo* (meaning "man"). Although the human race is varied in appearance, there are no other species of *Homo* alive today because all races of human beings interbreed successfully—and with an enthusiasm that confounds racists everywhere. Other species do exist within the genus *Homo*, but they are all extinct. Neanderthal man, *Homo neanderthalensis*, for example, is deemed to have been sufficiently different physically from us to warrant designation as a separate species—though it is now thought likely that, since we once inhabited the Earth at the same time, our two species could and therefore probably did interbreed successfully. As this example shows, it is hard to be definite about what is and is not a species, especially when some of them are extinct. But for all its difficulties, the species remains the most real thing in taxonomy.

In the animal kingdom, groupings higher than species and genus ascend through family, order, class, and phylum (the terms can be slightly different in botany, but the system is comparable). Thus the common mallard, *Anas platyrhynchos* of the Family *Anatidae* (ducks), belongs to the order *Anseriformes* (waterfowl including ducks, geese, swans) of the class *Aves* (birds) in the phylum *Chordata* (vertebrates and closely related invertebrates, all of which possess a hollow dorsal nerve chord at some stage in their lives) of the kingdom *Animalia*. The kingdoms (animals, plants, fungi, and so on) are also real in that it is hard to argue that a plant, for example, even if it apparently eats flies, is really anything other than a plant. However, even the kingdoms, and certainly all intermediate higher taxonomic groupings, between kingdom and species are, to some extent, imaginary. They are human constructs, the products of expert opinion, nowadays aided by computerized factor-analysis. It helps to think of this classificatory system in terms of building— and occasionally rebuilding—a house. Each species is a brick in the wall of the taxonomic edifice. Now and then, a new owner may decide to demolish the house and build a new one, sometimes with more rooms or more stories, sometimes with fewer; but they always re-use the old bricks. The species is sacrosanct.

Derek Ager, pioneering palaeoecologist and reluctant revolutionary of neocatastrophism, was also an expert in the classification of the small marine organisms called brachio-pods. He was suspicious of people who set much store by the middle taxonomic categories, like family and order. He had discovered, firsthand, how subjective they were, through work-ing on one of the very sources on which Sepkoski later relied: the 1965 Brachiopoda volumes of the massive, multi-volume *Treatise on Invertebrate Paleontology,* masterminded then by its founding editor, Professor Raymond Moore. One of the pur-poses of the *Treatise* was to bring some semblance of order to the classification of fossil organisms. As one of the many expert editors drafted in to perform this task on his beloved brachio-pods, Ager had noticed how some apparent extinction events, typically at the more important boundaries of the geological timescale, merely reflected the fact that scientists are territo-rial. All it took was for one splitter, a worker who tends to set up new species based on the slightest differences, or subdivide larger groupings into a plethora of lesser ones, to spend a career working in the Upper Devonian, for example, for there to be an apparent but wholly spurious extinction event at the beginning of the Carboniferous, where his activities ceased. This made Ager very cautious about computerizing range data, fearing the famous "garbage in, garbage out" problem. He was also suspicious of any paleontologist—and there were to be many more in years to come—who spent all their time in libraries and computer rooms and had no firsthand experience of defining actual species themselves. This was part of Ager's revolution-ary reluctance, and he was by no means alone. Many tradition-ally raised paleontologists saw Sepkoski's work as too removed from reality, and little above accountancy.

The advent of more rigorous methods in the way all cat-egories at all levels are assigned has since put taxonomy on a

much more objective footing, and greatly lessened the serious-
ness of Ager's reservations. And although this has been vital
in refining the Sepkoski database and its many successors over
the years, one is always drawn back to Dave Raup's cunning
observation about seeing patterns in dubious data. A massive
database containing data of mixed quality is not the same thing
as a dossier compiled by the intelligence services, for example,
consisting of small numbers of unreliable observations from
which wild inferences are then drawn by people looking for
the very things they end up discovering.

Paleontological databases consist of a huge and growing
mass of information, some dubious, some not, derived from the
published work of scientists who, whatever their other faults,
were completely innocent of the compiler's intentions. Should
any pattern emerge from such a compilation, and survive all
the statistical tests that can be thrown at it, the fact that the signal has come through, despite the noise, should persuade for,
rather than against, the reality of such a pattern. Nevertheless,
as with seeing meteors in the sky, deriving meaning from that
pattern—as opposed to merely recognizing and accepting it—
is a different process entirely.

✳

No matter how you cut it, there is no doubt that the diversity
of life has been increasing over the past half billion years; but
Sepkoski's graphs showed that it has not done so smoothly.
In a stock market, values of investments may go up or down
on short timescales but can be shown over longer periods to
display an overall trend—either up, usually during a bull market, or down, usually in a bear market. Similarly, if you had
been lucky enough to get in on the ground floor and buy stock

in the kingdom *Animalia* back in the early Cambrian when it was the latest thing but very risky, you would have done well; although along the way you would have had to endure a number of crashes that would have tested your nerve as an investor more than once. There have been five great mass extinctions—of which the end-Cretaceous extinction is one, though not the most severe—and the possibility that these crashes were periodic, and more reliably so than any stock market, brought Jack Sepkoski into our story earlier.

Despite the Alvarezes' discovery at the K-T Boundary, explaining why extinctions happen has not become easier because yet another potential killing method has entered nature's arsenal. Statistical analysis is incredibly useful in indicating what might be real, concealed within a mass of confusing data; but its results remain—like the shadows that flit across the crystal ball—open to interpretation. Statisticians can test to see whether trends are genuine, and not simply a product of the sampling method. But statistics do not, and cannot, give explanations. Explanations remain a matter for the human imagination.

So, for all that the apparent periodicity in mass extinctions that Sepkoski and Raup uncovered appears to be statistically solid, this is not really saying very much more than "this, which appears to be real, may (but only may) be really real." Until a convincing mechanism can be produced, such statistical chimeras remain tantalizing possibilities. Because extinction was in vogue at the time Sepkoski began publishing his compendious databases, it is not surprising that most attention focused on what his data revealed about the phenomenon, leading of course to those exciting but fruitless ideas about death stars and regular cometary visitations. However, death was not, thank goodness, the whole story. The data were also speaking about burgeoning life.

One of the most remarkable features of Sepkoski's curve was the way it showed that living things underwent a golden age of diversification from the mid-point of the Ordovician period. As more and more data were added, the database refined and extended to lower taxonomic levels than family, the more convincing the evidence for this mid-Ordovician bull market appeared to be. It even developed its own jaw-cracking name: the Great mid-Ordovician Biodiversification Event—or GOBE. The problem was, nobody could explain it.

The GOBE was quite different from the much better-known Cambrian Explosion, when in almost a geological instant animal life developed almost every major known body-plan (and hence all the major animal phyla) that we see around us today—as well as a few more that quickly died out. To call it an event seems to undervalue it; but it took place at or around the beginning of the Cambrian Period, 542 million years ago, and marked the first appearance of animals with hard parts—which therefore made them more easily preserved as fossils. The "invention" of hard skeletons revolutionized the fundamental ways in which animals could be constructed and earn their living. The GOBE by contrast was a flowering among existing forms of marine invertebrates—shellfish, essentially—which simply and inexplicably became more diverse in form, and therefore, one assumes, in the variety of ecological niches that they occupied.

After becoming established and growing sluggishly during the Cambrian Period, the animal life that swarmed in the shelf seas and topmost ocean waters of the Ordovician diversified at an exponential rate during the Period, establishing the population of animals that dominated the Earth's seas for the next 250 million years, at the close of which time the end-Permian mass extinction (biggest of the big five) led to its replacement by the fauna of the Mesozoic Era (in turn replaced by the fauna of our own era, 65 million years ago, when the dinosaurs met their nemesis). It was the most rapid rise in the diversity of life at any time in Earth history since the Cambrian, and—at the level of species and genus—by far the fastest. As many as 350 new taxonomic families were added to the roll call of life. And if biodiversity is plotted within the Ordovician Period, the most steeply rising part of the resulting curve is to be found—for most groups—just at the advent of the mid-Ordovician, 472 million years ago.

The Earth then would have looked very different from the planet we know today, as fragments from the collision that created both the Gefion Asteroid Family and the L chondrite meteorite clan began their long journey to Earth. Most land was concentrated in the southern hemisphere, and very little extended further north than 30 degrees; so that to us, used to a globe where almost the reverse situation applies, the Ordovician Earth looks "upside down."

Reconstructing the positions of ancient continents is difficult for such distant eras in Earth history, for two principal reasons. First, all the ocean floor that existed at that time has now been recycled, dragged back into the Earth's mantle by subduction (or in a few rare cases, become trapped in the vice of a mountain-building episode). Geologists can tell exactly how the Americas of today have drifted away from the opposite coastline of the Atlantic by the pattern of movement that is now written into the sea floor created to fill the space. Alas, no such helpful guide exists for the Ordovician continents.

The position of any point on the globe is fixed by reference to two coordinates—longitude (north–south) and latitude (east–west). To position Ordovician continents on the
globe, geologists search for the faintest remnants of contemporary magnetism preserved in volcanic rocks as they solid-
ify, or in certain sediments as their constituent grains (some
of which will be magnetic) settle out. These palaeomagnetic
measurements can tell geologists the latitude at which a rock
was formed, but not the longitude. For that, other evidence,
including the distribution of fossil species, comes into play,
though all these indications are open to interpretation—which
explains why all palaeogeographic reconstructions involve a
certain amount of artistic license.

A single mega-continent called Gondwanaland, comprising modern-day Africa, South America, Australia,
Antarctica, and Arabia, stretched from the southern ocean
to a few degrees north of the equator. Several smaller continental fragments stood some distance off. Nearest to the
Pole lay Baltica, comprising Scandinavia and much of northern Europe (west of the Ural Mountains, which had not yet
formed). Straddling the equator, Laurentia consisted mainly
of modern-day Canada/North America and Greenland.
Further east, at roughly comparable latitudes, lay another
continental island, comprising modern Siberia. Many smaller
island continents, including several slivers of Gondwanaland
that rifted off during the Ordovician Period and traversed
oceans to collide with opposing shores, included Avalonia, on
and around which the UK's Ordovician rocks were formed.
Other minor terranes (as geologists call them) included
Kazakhstania, whose name is self-explanatory, and a set of
fragments that eventually came together to form China, all

hovering offshore from eastern Gondwana at close to equatorial latitudes.

The atmosphere at that time contained only half the amount of oxygen that we breathe today, though there were high concentrations of carbon dioxide and water vapor, which together set up and maintained a "greenhouse" climate. As a result the tropics were much more widely spaced than they are today. The global ocean, which because of the prevailing greenhouse conditions was not subject to strong temperature differences between the equator and the pole, would have relied upon evaporation at mid-latitudes (and the resulting sinking of dense, saline water) to drive its deep circulation. This was much less efficient than today's thermohaline circulation; so much of the deep ocean—especially in the northern hemisphere—would have been completely starved of oxygen.

Global sea levels, relative to the continents, were much higher than they are today. This was especially true in the mid-Ordovician, when sea levels rose markedly and broad, shallow shelf seas extended widely over the continents. Many of these seas became restricted and hypersaline, giving rise to deposits of minerals that precipitate from concentrated solutions and are known as evaporites (principally gypsum and salt).

The prevailing climate over land areas like Baltica, close to whose northeastern shores the Swedish fossil meteorites would land, was warm, sunny, and arid. Very little life existed anywhere on land at this time, although primitive plants, including lichens and liverworts, were starting to colonize the new environment, together with some invertebrates. The very earliest spores, released by land plants, are found fossilized in rocks of the mid-Ordovician. In the ocean waters, floating organisms thrived, notably the fast-evolving graptolites, which drifted in the ocean surface currents and became quickly wide-spread across the world—enabling geologists to cross-correlate deep-water and shallow-water sediments of the same age from anywhere in the world. The extinct eel-like conodont, whose soft body usually left no traces but its distinctive spiky jaw apparatuses, performed a similar function, evolving quickly, distributing itself widely, and reinforcing the time-equivalence scale that geologists first established from the mid-nineteenth century onward using graptolites.

However, much of the rest of the Ordovician sea's fauna was provincial—which is to say that its constituent species had geographically restricted distributions, like many of today's marine fauna. Creatures like the brachiopods, for example, which were mostly fixed on or in the sea floor, together with sediment-grubbing bottom-dwellers like most trilobites, could not spread themselves across wide oceans because their larvae could only survive so long before finding a place to settle and metamorphose into their adult form. While this fact rules these fossils out as useful tools for long-range correlation of rocks, it makes them very useful in helping to constrain continental longitudes. For example, in sediments laid down around the margins of the continent Laurentia, the appearance of trilo-bite or brachiopod forms previously found only in the Baltic province, or vice versa, would suggest that the two continents had, by that time, become close enough for larvae to make the crossing.

Most intelligent among the life-forms in the Ordovician sea were probably the free-swimming cephalopods, then as now the cleverest invertebrate creatures by far. Other molluscs included the bottom-dwelling snails, and clams—at that time as subordinate to the brachiopods as brachiopods are to clams today. The Ordovician was an especially successful time for crinoids (sea lilies) and their cousins the blastoids—all distant relatives of today's urchins and starfishes. Early vertebrates, in the form of jawless armored fish, cruised the Earth's waters for the first time. Ordovician shelf seas were generally poor in reefs, although those that did form consisted mostly of tiny colonial animals called bryozoans, sponges, and algae. Now-extinct types of coral were also present and, from the middle Ordovician, the stromatoporoids began the reef-building career that was to blossom through the ensuing Silurian and Devonian Periods.

Toward the end of the Ordovician, as Gondwana drifted south across the pole, the carbon dioxide content of the atmosphere fell, and the global greenhouse gave way quite abruptly to an icehouse world. The icehouse is Earth's other stable climatic state—the one in which we find ourselves today, with permanent ice at both poles, frequent glaciation, and narrow tropics. Glaciers began to become widespread, in a catastrophic cold snap that seems to have lasted for about a million years. As the Ordovician ended, life's greatest ever explosive diversification was brought to an end by a major mass extinction. About 60 percent of animal genera and 85 percent of all species vanished—making it the second most deadly mass extinction in Earth history, due probably to the combined effect of falling temperature and lowering sea level. (Ice ages always cause falls in sea level, for the obvious reason that more of the world's water becomes piled up on land). But that disaster lay far in the future when the first debris from the Gefion collision began to rain on the great parade of burgeoning Ordovician life, many of whose shallow marine species were shellfish called brachiopods.

Paleontologists and geologists generally are much more familiar with the Phylum Brachiopoda than most biologists for the simple reason that although not totally extinct, somewhere around 99 percent of all known brachiopod species are no longer with us. Brachiopods' glory days, in other words, are well behind them.

Looking superficially like clams, with two hinged valves enclosing their filter-feeding bodies, brachiopods originated in the Cambrian Period, and dominated the sea floors of the Earth for about 400 million years. Some were attached by a kind of tether, like modern mussels. Some were cemented, rather like oysters; many more lived part-buried in sediment like cockles, or lay loose on the sea floor like scallops. They invaded ecological niches worldwide, at different depths and energies of water; but as these comparisons show, in all their typical habitats, they are now supplanted by the clams, which took over from them after the end-Cretaceous extinctions, at which time brachiopods suffered a hammer blow from which they never really recovered.

The name brachiopod means "arm-foot" (in Swedish they are known as *Armfotingar*)—a name that refers to the large organ, covered in tiny waving hairs, that they all carry within their shells and use for breathing and food-gathering. Derek Ager believed that all paleontologists—especially, perhaps, if they wished to earn the right to become palaeoecologists— should become experts in at least one taxonomic group first; and for him it was the brachs, and the Order Rhynchonellida in particular. When I once, on a student field excursion to Sardinia, called out that I had found a rhynchonellid, Ager rushed over with the slightly worrying cry of "Be careful what you say—you are speaking of the animals I love!" Those outside the charmed circle can never understand the peculiar affection that a specialist feels for his or her preferred group of beasts; and if this—and such minor foibles as the fact that Ager called his yacht the *Rhynchonella*—seems obsessive, and more than a little eccentric, so be it.

Despite the near-invisibility of brachiopods in the seas of today, their importance in the fossil record is second to none, as is the number of volumes devoted to describing them in the monumental *Treatise*. The revised part H—descendant of the 1965 first edition on which Ager worked—runs now to six separate and rather mighty volumes over a foot thick, and the modern classification that they established formed the basis of a study of Ordovician brachiopods carried out as part of a three-year international research program to learn more about the Great Ordovician Biodiversification Event.

This survey clearly revealed that the GOBE gave the brachiopods their big break. Their origins in the early Cambrian apart, the most important single event in the whole long history of the phylum Brachiopoda was the unprecedented development during the mid-Ordovician. By the end of the Period, all but four of the many innovations that appeared during the entire evolutionary history of the brachiopod shell had come into play. What is more, when the diversity of brachiopods, expressed as the number of orders and of genera within each order, is plotted against a timescale for the Ordovician, brachiopod experts found that diversity did not in fact rise gradually but shot up at the base of the mid-Ordovician—at precisely the moment when the first L chondrite meteorites from the breakup of the Gefion family began to strike the Earth.

Meanwhile, in Sweden, to refine the resolution of diversity curves covering the bombardment, Birger Schmitz assembled another international team to conduct the most detailed, bed-by-bed sampling of brachiopod fossils ever undertaken for the mid-Ordovician anywhere in the world. In all they sampled eight overlapping sections in Baltoscandia, collecting over 30,000 fossil specimens. These were then identified and the data, at the level of species, plotted on a graph, showing time on the horizontal axis and number of species along the vertical. The researchers also plotted individual extinction and speciation events, the points at which new species first appear in, and disappear from, the section. They found that by far the largest changeover in brachiopod species occurred not just roughly but *precisely* at the same horizon where the flux of L chondrite material reached its early peak.

The coincidence of the two events is spectacular and undeniable. As ever, though, the question scientists must ask is one of meaning. It is a sacred rule of statistical interpretation that if two variables change together, this need not necessarily imply a direct causal relationship. The catchphrase runs "correlation is not causation." Furthermore, the name Great Ordovician Biodiversification Event does it a disservice. Only when viewed from a great distance with a small timescale does this massive burgeoning of life through the Ordovician look anything like an event. Closer inspection, greater familiarity with the detail, makes it inconceivable that one could ascribe all of this diversification, over hundreds of millions of years, to a single cause. To do so would be to fall into the same trap as the disciples of the K-T impact theory in its purest form.

The GOBE was unprecedented and stands unmatched in the history of life. There have been other great diversifications, but always as recovery phases following mass extinctions. The GOBE was deep and far-reaching, operating at species and phylum level. A comprehensive answer to the question of what caused the GOBE is not going to pinpoint one factor alone, just as the K-T extinctions cannot really be attributed to the impact of one asteroid. A rain of meteorites may indeed have somehow stimulated a sudden diversification in brachiopod species at 470 million years. But this does not mean that other causes were also not coincidentally operating, before and afterward.

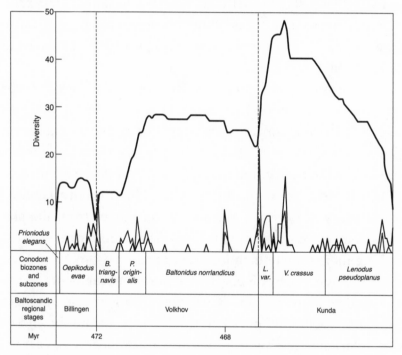

The diversity of species of brachiopod fossils through part of the Lower and Middle Ordovician in Baltoscandia. Note the dramatic increase in biodiversity (top line) occurs just at the base of the Lenodus variabilis zone, and is matched by two simultaneous spikes in numbers of extinctions and numbers of first appearances of new species (lower lines).

The wide extent of shallow shelf seas at equatorial and mid latitudes, also undoubtedly played a part. That there were at least three cycles of marine transgression and regression; that there was a changeover from greenhouse to icehouse that prepared the way for the end-Ordovician glaciations; the widely spaced continents, the repeated rifting of many fragments from the edges of the megacontinent Gondwanaland, on which geographical isolation could then work to produce new and distinct populations—these factors did not cease to be important just because the world was suddenly bombarded by rocks from space. Ideas do not become wrong just because they grow old—an error we have made too often in the past to make again.

This multi-cause approach, though operating here in favor of glorious life, instead of its nemesis dull death, is Derek Ager's "silly diagram" in operation once again. Big causes on long wavelengths mix with small causes on short wavelengths, and create their effect only when they fall into phase. It may well be that change, in the form of a great biodiversity increase, was already afoot when the shower of stones from the sky added one more stimulus to the mix. This would explain why, looking more widely than just the brachiopods, many groups seem to have begun diversifying before the fragments of the L chondrite body first made earthfall and just as many taxa went into apparently prescient decline long before the end of the Cretaceous. The mid-Ordovician does appear to have been the sort of period to which the supposed Chinese curse—"May you live in interesting times"—might have applied.

The idea of an increased mid-Ordovician meteorite flux is now, thanks to the work of Birger Schmitz and his co-workers, firmly established and is currently generating interesting new findings. Professor John Parnell of the University of Aberdeen has recently surveyed the published data and discovered that mid-Ordovician sequences all around the world contain a much higher than average quota of massive, chaotic slump deposits called megabreccias.

These rather terrifying sediments, also known as olistostromes, are created by submarine landslides. At the edges of continental shelves, sediments build up to enormous thicknesses and become unstable. A trigger event, such as an earthquake, can then send tens or hundreds of cubic miles of poorly consolidated sediment cascading down into the ocean deeps. The resulting mass is a completely chaotic deposit, with individual blocks ranging from yards across to—literally—the size of mountains. If you imagine allowing a sherry sponge trifle to slide out of its bowl, over a washboard and into the kitchen sink, you will get a good small-scale picture of what a megabreccia can be like. A sudden submarine landslide can itself give rise to tsunami waves, such as devastated the northeast coast of Scotland just over 6,000 years ago when a section of Norway's continental shelf gave way.

Birger Schmitz acknowledges that between five and ten times larger impact craters date to the mid-Ordovician than one would expect under normal conditions. Therefore, during the 10 million years of the GOBE, the Earth could have received one "Chicxulub-sized" impact—6 miles across—or perhaps two of 3 miles diameter. So Parnell was not surprised to find that all over the world, rock sequences laid down on the continental margins between 470 and 460 million years ago show ample evidence of catastrophic submarine landslides.

From the limestone margins of Baltica to the Pacific side of Laurentia (now the Yukon); from the northern margin of Avalonia (now the Lake District and the Isle of Man) to north China and South Korea, olistostromes can be found, often containing blocks of shallow-water sediments over half a mile across that foundered into the ancient deeps. Even volcanic island arcs, sitting out in what was then the Iapetus Ocean (between Avalonia, Baltica, and Laurentia) were not immune from these huge slides. Near Trondheim, Norway, a mid-Ordovician megabreccia containing limestone blocks over 109 cubic yards in volume can be seen to pass laterally into volcanic rocks that were once erupted on such an island arc, situated originally somewhere offshore from the margins of Laurentia.

The clear implication that meteorite flux increased at the very least five to ten times above modern levels during the mid-Ordovician—and yet have had a stimulating effect on the sea-floor life of the time—is particularly interesting when compared to the situation at the end of the Cretaceous. We know that there was one huge impact then, and maybe, if Gerta Keller is right about the age of Chicxulub, more than one. Her interpretation is hotly disputed, but it is certainly not impossible. As the mid-Ordovician clearly demonstrates, impactors can and do arrive in swarms. What is more, given different circumstances (a rich mid-Ordovician planet burgeoning with life, as opposed to a sick, end-Cretaceous planet dying from the noxious volcanic pollution) impacts might then do living things a great deal more good than harm.

Attempting to understand how they might have achieved this will, I hope, convince anyone inclined to doubt it that life is indeed infinitely more interesting than death.

10

UNREST CURE

You've heard of Rest-cures . . . well, you're suffering from overmuch repose and placidity, and you need the opposite kind of treatment.

SAKI (H. H. MUNRO)

Scientists spend a lot of their time simply mapping what is out there. Astronomers scan the heavens for exoplanets, stars, nebulas, supernovas. Geologists map rocks, paleontologists find and catalog fossils, biologists name living species. Vital though this work is, the greatest feats of scientific understanding come next, usually when someone looks at their chemical elements, stars, strata, or species and sees them not as objects in beautiful isolation, but linked by a historical process. Each object becomes the cut end of a skein of threads, stretching back through time to the beginning of things.

Each individual element, species, or stratum becomes then the product of an evolutionary process: from the earliest stars to our own Sun; from the rarefied hydrogen and helium of the Big Bang to the heavier elements that build the Earth and life's muddy vesture of decay; from the living species to the earliest sparks of life; from a tangled mass of deformed rocks to plate tectonic processes that unite all Earth science through the endless cycle of forming and re-forming supercontinents. Knowing "what" is transfigured in a flash into understanding "how." What were merely objects in a cabinet of curiosities become the outcomes of process—and processes spring from law, as surely as now springs from then.

The child, as Wordsworth pointed out, always becomes father to the man, just as all things in nature are the products of processes that snake back into deep time. And although the process of nucleosynthesis in evolving stars proceeds inexorably by the laws of physics, most other processes, such as biological and geological evolution, are the product of historical events that cannot have been foreseen and might not necessarily have happened in the way they did. The state of things now is contingent upon what happened in the past, and while it might be hard to imagine things otherwise, the sense of inevitability that they give us is almost entirely illusory. In the complex web of history, the great effects are reserved for those moments when myriad influences accidentally pull together one way or the other, as Derek Ager was so fond of reminding us. What this in turn means is that the effect of a particular agent on subsequent history will depend crucially upon what else is happening at the time—the context of the event will give it meaning.

In order to understand how a massive meteorite strike in the late Cretaceous can have the opposite effect from raised meteorite flux in the mid-Ordovician, it is important to introduce a little ecological theory. Ecology charts the way organisms in communities affect and are affected by their physico-chemical environment, and on long timescales what this means is the interaction of biology with geology. The simplest way to explain this is to think about a piece of landscape that has changed—in geological terms—very rapidly. I shall try to conjure up such a place in your imagination by describing somewhere from the landscape of my own youth, the sweep of coast between Great Tor and Oxwich Point in the Gower Peninsula of South Wales.

From a certain vantage point on a high and little-traveled path near Nicholaston, you can look out over a wide expanse of dunes stretching from the base of an ancient cliff-line to the present high-water mark. You stand on the remnants of gray limestone cliffs, first created 10,000 years ago, after the glaciers melted for the last time and the sea rose 273 yards to its current level. Those old cliffs, which have not felt the crash of waves for millennia, are now covered by briars and ivy, pennywort, primroses, and forest. They are, in a sense, fossil cliffs; and the tell-tale shapes, carved into them by those long-retreated waves, do not lie. Where dunes now touch them, hundreds of yards from where the highest spring tides now reach, wave-smoothed overhangs bear witness.

The dunes formed as wind-blown sand, torn from the beach by the south-westerlies roaring up the Bristol Channel from the Atlantic, and dumped at the cliffs' feet, pushing back the sea and damming the Oxwich River to form extensive fens. Biology and geology started working together. The dunes were quickly stabilized by the underground stems (rhizomes) of the spiny marram grass, *Ammophila arenaria*. As a thin soil developed, more species came in: *Cakile maritima,* the sea rocket; *Eryngium maritimum,* the sea holly, once favored as an aphrodisiac; *Festuca rubra,* the red fescue. Eventually, thousands of years after the process began, light deciduous woodland developed—a multi-layered, diverse community housing thousands of species.

Ecology is still a rather young science and one that in many areas is still searching for generally accepted principles governing the way communities alter in response to the different pressures affecting them. This is not surprising. Living things are intrinsically more complex than objects hurtling through space, obeying the simple dictates of physics. Not only are living things themselves complicated; communities are more complex still, and they respond over time to both internal dynamics and external forces that themselves are subject to change often in complex ways that we do not fully understand. Complexity piles upon complexity, and uncertainty upon uncertainty, at every turn.

It is fair, though, to say that deciduous woodland, the climax community in our example, contains more species than the dune field. The climax community is more diverse; undisturbed natural succession has produced greater complexity. Similarly, within a few millennia of the end of the Ice Age, the British Isles, separated once again from Europe, became quickly clothed in thick forest—the original greenwood—of which almost nothing now remains. Was that the golden age of biodiversity, after which we have seen nothing but change and decay all about, largely at our hands? Complexity and diversity are the hallmarks of the healthiest ecosystems, so humanity today is deeply affected by issues of biodiversity and how best to foster it. If the example of ecological succession from dune to forest were our only guide, then we might think that all humans need do is leave well enough alone. Such is the belief among many environmentalists, often stronger on emotion than understanding, who tend to oppose any attempt to exploit the Earth's resources on precisely these grounds. But as a species we have no option in this matter. When it comes to exploiting the Earth, ours cannot be a simple, binary choice since either option—exploiting the planet to exhaustion or not exploiting it at all—leads to extinction. Our only rational course is to use our understanding of nature to devise a sustainable future that will allow us to use the planet's resources in such a way that does not leave our descendants homeless.

Happily, what ecological science teaches us is that environmental disturbance is not only commonplace in nature—it maintains and develops biodiversity. That is why it is so important to understand what the end-Cretaceous extinction and the Ordovician meteorite storm are telling us.

Behind where I grew up on the outskirts of Swansea, South Wales, stretches a scarp-and-vale topography formed on the northerly dipping rocks in the southern limb of the great geological downfold that is the South Wales coalfield. Thick sandstones form the hills, and soft shales with coal the intervening valleys. When I was a boy, these hills were a kind of no-man's-land. Although fields had been marked out on them centuries before, they were then, as now, largely unfarmed. Though the fields had often been sold, they were not yet built upon. Although there were mine workings, quarries, and waste heaps, the vestiges of these nineteenth-century industries were melting back into the land and becoming archaeology. It was a place no humans seemed to want; but life never abandons an area.

These Pennant Hills, which I name after the thick sandstones that underlie them, bore a rich soil that had been established by a light deciduous forest on a clayey, bouldery subsoil called glacial till—the rubbish left by retreating ice 10,000 years ago. That process itself would have involved a distinctive ecological succession as plants and animals came in to fill the newly exposed territory that had lain for millennia under the smothering ice. Light forest became established and remained until perhaps a few hundred years ago, when the first agriculturalists built farmsteads and cleared the forest for grazing.

The field boundaries, much collapsed when I was a boy, probably dated from the Enclosure Acts, passed from the mid-nineteenth century onward, not long before the onset of a half-century-long agricultural depression (from about 1870) which saw them abandoned. Many farms became derelict, suffering the same fate as the extractive industries. Throughout the nineteenth century, struggling agriculture had vied for control of the Pennant Hills with mining, quarrying, and brickmaking. Shallow workings were opened up to exploit the thin, impersistent coals. Small quarries were cut for building stone, creating man-made cliff habitats that mimicked those other artificial crags among the wrecked, ivied gable-ends of rotting engine houses and chimney stacks. Ecological succession repeated itself as *Betula*, the silver birch, recolonized the heaps of blue-black shale, abandoned beside the worked-out adits—just as they had done thousands of years before, when the glaciers first retreated and lichens and mosses had afforded their seeds a first footing on the barren periglacial outwash.

As the farms failed, their un-mown fields ran wild, becoming meadows rich in annual flowers and insects. Tough bracken (*Pteridium aquilinum*) with its deep rhizomes began to spread from the hedgerows, together with the last remnants of the original woodland—*Quercus*, the oak, and *Fraxinus*, the ash; hazel (*Corylus*) and *Ilex*, the holly. Around the bracken's edges, gorse bushes (*Ulex*) sprang up. For a hundred years or so, a landscape in dynamic balance, with slowly re-advancing forest, spreading bracken, retreating grassland and gorse bushes, became established on the Pennant Hills.

Wildfires, which became more frequent as the town encroached, kept the forest back; but it did not much affect the bracken, or the aptly named fireweed (rosebay willowherb,

Epilobium angustifolium) whose rhizomes also live safely under-ground, ready to sprout afresh in spring, suppressing other plants by shading them and, in the case of bracken, by secret-ing natural poison. Gradually the hills became dominated for the most part by these two fire-resistant species; but all this changed one year, when a gas pipeline came through, its trench cutting over the crest of the hills.

In the clinging mud of a Welsh spring, heavy machinery pushed a broad swathe through the bracken, 33 yards wide. The rhizomes were smashed and churned; and when the next spring came, delicate fiddleheads failed to nudge the clods aside. In their place appeared hundreds of thousands of seed-lings of the spiny gorse. Already present in the soil, dispersed over many years by its exploding pods, gorse was able to establish complete dominance in one season. After two years, it had become an impenetrable thicket that turned in summer into a vivid yellow stripe down the hillside, a peppery coconut scent filling the air, yellowhammers perching on their topmost twigs as the summer sun dried their seed pods and shot out cannonades of tiny black grains. The very same ecological phe-nomenon had produced the carpet of poppies in Flanders after the devastation wreaked by the trench warfare of World War I.

Years went by. Fires passed through, and the gorse burned bright and hot. New growth sometimes sprang from some gnarled, charred bases, but its glory days were over. I returned there recently to see how things had changed. Now, decades later, except for a few old timers still keeping their necks above the competition, the steady army of *Pteridium aquilinum* has once more reclaimed the hill.

This story shows how different plants respond differently to environmental disturbance. Bracken employs a strategy based upon the occupation of space. Once in, such species dominate completely to the exclusion of others. However, it takes them time to do this. If dominance should be catastrophically broken, the setback lasts for years, and with bracken out of the way other species have a chance. These species have effective seed-dispersal mechanisms and grow quickly, making hay while the sun shines on the exposed soil. They are the carpetbaggers of the living world—opportunists, always eager to exploit whatever chance throws their way.

Ecologists have sadly chosen to name these two strategies after two factors in an algebraic equation—giving us "r" (for the opportunists) and "K" for slow-growing long-term occupiers. Environments where K-strategists hold sway will, in ecological parlance, become climax communities (or very stable stages in the slow progress toward a climax) because once their stranglehold is established there is almost nowhere new for the community to go. Without the disturbance that gives other species a chance, a low-diversity climax may hold sway. Leaving well enough alone is not always the best way to foster biodiversity.

As a boy ecologist, I came to know these Pennant Hills intimately. In fact, at the age of about twelve (when I should have been out playing football and breaking windows) I would regularly disappear with a quadrat and notebook to do ecological surveys of the sort I would not do again until university. By the time I reached the age that Elagabalus was when he became emperor of Rome, I knew my few square miles so well that I regarded it as my territory. Over the years, some of it became subsumed beneath 1970s housing of varying degrees of shoddiness. Today, some of that land so long left fallow is going back under the plough for the first time in perhaps a century.

Change remains the only constant; but I had taken a valuable idea from my apprenticeship. I had learned that the unrest cure, environmental disturbance, breaks the grip of ortho-doxy—wiping out the ruling occupiers of a particular piece of life-space—just as effectively as new ideas sweep away old, revolutionizing our understanding.

Happily such youthful studiousness did not extend so far as reading scientific journals, so I was not aware at that tender age of the distant thunder of contemporary ecologi-cal theory. However, had I heard it, I would have discovered that the idea of a positive correlation between certain levels of environmental disturbance and diversity had, like almost all thoughts, already occurred to others. It is not a difficult idea; I had intuited it in my early teens. But ideas, like meteorite strikes, depend for their effect upon the historical context in which they arrive and are interpreted. Both must occur at the right time, to the right mind.

Most textbooks of ecology will tell you that intermediate disturbance theory, as it is called, dates from 1978, and a paper written by J. H. Connell in *Science*. However, this is a textbook fact parroted by subsequent authors who copy their predeces-sors without consulting original sources. Connell's mention of it, in a paper published in a prominent journal, merely made intermediate disturbance theory better known. The theory had in fact been part of the intellectual toolbox of ecologists since the 1940s. What is more, it would soon enjoy an unprecedented vogue. By another of those chances that come to seem like predestination because of the way they affected subsequent events, environmental disturbance and its effect on biodiver-sity confronted me forcibly in my doctoral research among the fossil reefs of Gotland.

In modern seas, tropical reefs are the most biologically diverse communities known to science. They are mostly built by a type of coral known as scleractinians, which have come to dominate since first appearing in the middle Triassic, about 240 million years ago. Reefs are rather unevenly distributed through the fossil record, and they have certainly not always been coralline. In the late Precambrian, algae built them. In the Cambrian Period, frame-building creatures called archaeocyathids were their prime architects. In later times they became dominated by other groups—such as the now-extinct coral orders (tabulates and rugosans); stromatoporoids, rudists (aberrant oyster-clams that thought they were organ pipes), and so on. But whichever group was principally involved, a tendency toward the reefal habit has been a recurrent feature of marine life throughout its history.

The Silurian patch-reefs of the Visby Formation grew to a few yards across at most, initially (at the base of the unit) in deep, muddy water. Sea level was gradually falling at that time, so the nearer one of these fossil reefs was to the top of the Formation, the younger it was, the shallower the water in which it grew, the more diverse its biology, and the more dominated it was by corals and stromatoporoids with robust, massive or plate-like growth-forms. This ecological change was a response to greater water energy and illumination. But such adjustments paled into insignificance beside a sudden transformation that occurred right at the top of the unit.

Here, where my rocks gave way to those of the unit above—the Högklint Limestone Formation, being abseiled by my agile climbing colleague Nigel Watts—everything changed. Seemingly in a flash, my smallish patch-reefs flowered into huge, towering edifices, hundreds of yards across and tens of yards tall. They developed complex internal structure with internal cavities. The sediment between colonies became packed with algae and cemented, creating massive reef rock, thriving in the light of the shallows and frequently eroded as storm waves scarred the emergent structures.

The Visby reefs had not only literally found the sun. By crossing wave-base, the deepest point reached by the turbulence of storms, they had felt the beneficial effects of the unrest cure. New ecological niches, opportunities for different modes of life, had opened up; not only on the reefs' wave-smashed exterior, but in the silent gloom of their interior cavities. Such caverns became home to large numbers of specialized cavity-dwelling organisms—creatures whose fossil remains I had found in older rocks but only in small numbers, occupying tiny shelter cavities beneath individual corals or stromatoporoids sitting on the quiet sea floor of the deeps. Now, all these different species, with their widely divergent environmental demands, were coexisting in one place in much greater numbers—all thanks to environmental disturbance.

Environmental disturbance is very sensitive to scale and degree. To make an analogy with economics, governments regulate free markets to preserve a diversity of suppliers, large and small; they introduce rules to prevent any single company creating a monopoly. But just as regulations must not be too severe and risk choking trade altogether, environmental disturbance is only stimulating to biodiversity if it is neither too severe, nor too frequent. If either frequency or severity increases beyond certain levels, the ecological succession never moves off first base, and the equivalent of the gorse-bush forest will persist forever.

The mid-Ordovician world into which the debris of the Gefion break-up event came showering down was one ideally suited to the diversification of life. Global biodiversity was low, by modern standards, but because sea levels were high there was plenty of habitable space on the broad shallow shelf seas. The continents also were quite widely dispersed, which not only meant a long global coastline, but separated different gene pools from each other. When specific gene pools become isolated, they pursue their own evolutionary destinies. Thus biogeographical diversity encourages species to split from each other, stimulating evolution to produce more species through time.

The fauna of the shelf seas at this time was largely dominated by two broad groups of creatures—those that grubbed around in the sediment, and static, filter-feeding creatures like brachiopods that were fixed to one place, occupying space. The same ecological rules apply to fixed animals as apply to rooted plants; certain species will eventually reach a position of natural dominance in certain environments after the completion of an ecological succession. Thereafter, they will form distinctive communities according to whatever environmental parameter is most important. In shelf sea communities, that tends to be water depth.

What Birger Schmitz's discoveries in the mid-Ordovician suggest is that the peppering of the Earth with impacts, one of which at least could statistically have been "Chicxulub-sized," acted like fire on a bracken-covered hillside, or a storm in a pine forest, or wave action on a coral reef. Large areas of the Earth's continental shelf would have been affected. Dominant species will have been wiped out locally, just as Ernst Öpik envisaged in his limited-theater cometary strikes, erasing the kangaroo from Australia but allowing other marsupials in other areas to survive. More rapidly colonizing species would then move into space previously closed to them. Given a chance to become established, biodiversity would have been further boosted, resulting in the more rapid evolutionary speciation that we know today as the Great Ordovician Biodiversity Event.

The idea that meteorites might bring life to Earth is not new and, literally interpreted, is still looking unlikely. Yet as Schmitz's discoveries show, interpreting the idea more metaphorically might indeed make it true. What idle God, casting stones into Huxley's creation cockshy, could have predicted that bombarding shallow Ordovician seas with sub-lethal meteorite impacts for a million years and more would be just what the doctor ordered for stimulating evolutionary diversification?

Science surprises us because truth—which is to say, nature—is stranger than any theory we can dream up about it. Many have tried to express it. Sir Arthur Eddington said it well, but so did J. B. S. Haldane, who put it slightly more pungently when he wrote: "My own suspicion is that the Universe is not only queerer than we suppose, but queerer than we *can* suppose."

EPILOGUE

Fear no more the lightning flash, Nor the all-dreaded thunder-stone

SHAKESPEARE, *CYMBELINE*

The human species is the ultimate opportunist, and like any r-strategist species, humans are roaring through the Earth's resources at a rate that cannot be sustained. Like all others of our kind, we will one day hit the wall unless, unlike those others, we use our knowledge and turn our behavior into something more akin to that of a K-strategist, before it's too late. Going the way of the dinosaurs—succumbing under the onslaught of environmental catastrophe during a mass extinction—would be a particular irony for us. We would be the first species to pursue the r-strategy to an extent that affected not one area of planet, but the whole globe. Unlike other victims of mass extinctions, the one happening all around us today is one we are bringing upon ourselves, even while we are aware that we are doing it, and of where it will lead if we don't stop. How sad a fate for the first and perhaps last example of a sentient being to evolve on the heedless Earth.

People have always interrogated meteorites for meaning. Since they first fell on professors in Siena during the age of reason in the late 1700s, when rational thought was in vogue and superstition on the wane, posing the question how and understanding them as products of history has proved the best guide to what they might mean for our future. They have also taught us that exactly what role they perform in the evolving story of Earth history is dependent upon the Earth's place in that history—in much the same way that an observer's reactions to witnessing a meteorite fall depend upon the witness's social, historical, and economic context, or the way that certain scientific ideas flashing through the firmament of a human mind may flourish or wither as much according to contemporary politics, for example, as contemporary science. The realization that the dinosaurs' extinction was at least affected, if not entirely effected, by a meteorite forces us as a species to re-assess the risks that face us. That exercise has revealed the extent to which the overriding danger to our survival—ourselves—vastly outstrips any danger posed by space debris.

This is perhaps best explained by calling the odds on that one occasion when we know for sure that a mass extinction was at least partially brought about by the arrival of a major impactor—the end-Cretaceous extinction.

One hundred and sixty million years ago, before the dinosaurs had evolved, a random collision somewhere out in the Asteroid Belt, possibly the one that destroyed the parent body of the Baptistina Asteroid Family, sent a flurry of objects into resonant orbits, from which many were propelled toward the inner Solar System by Jupiter's gravity. For the next 93 million years, during which time dinosaurs evolved into one of the most successful groups of vertebrates ever to walk the Earth, one errant 6-mile fragment harmlessly crossed the Earth's orbit at moments when our planet was elsewhere.

If dinosaurs had become just a little more intelligent and had begun scanning the heavens for Earth-crossing asteroids, they would eventually have seen their nemesis. Using their knowledge of planetary clockwork, these dinostronomers would have been able to predict that the tumbling fragment would one day experience a very close shave with Earth, far in the future. They would not know that by that time they and their kind would already find themselves in the throes of extinction, along with much else that was alive in the late Cretaceous, thanks to many different environmental factors, including an outburst of volcanic activity of almost unimaginable proportions.

The story of subsequent events has become a familiar one. One fine day—and Derek Ager would have suggested a Tuesday afternoon because that was always his joke—it would finally hurtle down Earth's gravity well and become the last straw for *T. rex*. But this space rock would land not on a rich, burgeoning Earth. It would discover one where life itself, although lately seeming to be proceeding successfully, was now being poisoned by hundreds of thousands of cubic yards of basalt being erupted in the Deccan Traps of India. Ten trillion tons of sulfur dioxide, a similar amount of carbon dioxide, not to mention a hundred billion tons of gases including hydrochloric acid, would already have been injected into the atmosphere, in (as new and as yet unpublished research is revealing) perhaps as little as a few tens of thousands of years. No other terrestrial phenomenon correlates so positively with mass extinction events as the eruption of these so-called Large Igneous Provinces. The coincidence of a 6-mile-diameter impactor with such a catastrophe would—at the very least—have been very different from a similar event occurring in the balmy mid-Ordovician. Same agent, different result. In an age when we are supposed to understand the unpredictable consequences of actions within complex and chaotic systems, this possibility should not seem too surprising.

A coincidence of several unfavorable trends is bound to occur now and then in the long history of our planet—as normally out-of-phase cycles suddenly swing together and conspire to wipe the smile of life from the face of the Earth. Yet even when such an unthinkable disaster occurs, it may not be bad news all round. History is famously written by winners, who then tend to assume that the events that led to their success were in some way historically inevitable. Among those winners in macroevolution's game of chance, we must number our own ancestors. And while it should never be adopted as a reason for indulging in them, even the most terrible wars and holocausts have their seminal survivors, the ones who escaped in time to thrive elsewhere—just as in the horrific litany of waste and death that is nature's way.

So, as we now threaten ourselves with extinction, should we really be worried about meteorites any more? The Sun sits at the center of a bird's nest woven from the intersecting orbits of thousands of similar asteroids, many with the potential to hit the Earth. Yet in the vastness of space our planet is a tiny object. Even if the orbit of such an asteroid crosses our planet's path, for there to be a collision both must be at the right points in their orbits at the same time. But what does "at the same time" really mean? In space, things move at tens of miles per second. At such speeds, it is a lot easier for two objects to miss than to hit. How long would Earth and asteroid have remained in gravitational range of one another? What was the difference in time between a catastrophic hit that changed history forever and yet another near miss among millions?

It turns out that the time interval outside which Earth and asteroid would have gone on their separate ways was seven and a half minutes. If the asteroid that had *T. rex*'s name written on it had been running just a little earlier or later on that fateful Tuesday afternoon 65 million years ago when the Cretaceous Period came to an end, then Earth and asteroid would have grazed by one another harmlessly. Putting this in perspective, out of the 4.567 billion years that have elapsed since our planet formed from the protoplanetary nebula, only 450 seconds made the difference between our modern world, ruled (as we like to think) by us, and an alternate reality where the dinosaurs never died and our timid ancestors stayed home, quivering in their wormy burrows. At least, that is how one would put it if one really believed that the K-T extinction was entirely caused by one devastating impact and nothing else.

In 1980, when Luis and Walter Alvarez and their co-workers revealed that the end of the Cretaceous was marked all over the world by a thin layer rich in the rare element iridium, their conclusion was electrifying partly because, as all geologists know, if something has happened before, given time, it will probably happen again. The Alvarez idea imposed the need for action, and astronomers were eventually charged with evaluating the danger of a repeat performance. If we can see it coming long enough in advance, we can compute an orbit and work out when the asteroid will most likely hit. Armed with that information, we can envisage solutions—by deflecting the incoming missile off its deadly trajectory—or, in Hollywood's preferred, but probably most unwise scenario, blowing it up.

In 1991 an international working group sponsored by NASA was set up. As we have seen, NASA made a nod to the great Arthur C. Clarke by calling the resulting research project Spaceguard, and its main aim would be to identify all or a specified proportion of the known Near Earth Objects with diameters of more than 1 or 2 miles. The project began in earnest in 1998, aiming to find 90 percent of these objects within a decade. By June 2008 Spaceguard's astronomers had found 742 objects—probably, they figure, about 79 percent of the total. Of those discovered, not one was found to have more than a negligible chance of hitting Earth in the next fifty years. This fact alone had a remarkable effect on the statistics of risk. As a result of Spaceguard's mapping work alone, scientists were able to reduce the posted global risk of death by extraterrestrial impact to a negligible 1 in 720,000. It is interesting to compare this figure with some other unpleasant ways to go that nature—and our fellow humans—might visit upon us.

First of all let us recall that although meteorites have taken out a few vehicles, outside myth and dreams no humans have ever been killed by them, and, as we have seen, even the unfortunate Nakhla dog is probably a myth. (Though meteoriticists maintain a sneaking fondness for the creature. They like to joke that if this tale were true, then it would be the only example of an earthling ever being killed by a Martian.) More seriously, the average global risk for all living people of dying in an earthquake stands at about 1 in 130,000. The global risk of dying in an airplane accident—including for people who will never set foot in an airport—is 1 in 30,000. However, the global odds of dying in an automobile accident are three orders of magnitude greater, at 1 in 90—and remember, that also goes for every member of the remotest tribe in Papua New Guinea, who will never even see a Chevy Malibu. So, if by contrast you are a commercial traveler working out of Detroit, your actual personal risk of death by car is—astronomical; yet you and your insurance company readily accept it every day.

Thanks to Spaceguard, we already know that our chances of dying as a result of something hitting us from space stand on a par with death by firework (about 1 in 600,000). I am writing this in Nancy, France, on Bastille Day, during the 72nd annual conference of the Meteoritical Society. For this reason I feel the chances of death by firework more keenly than I would normally. Nevertheless, I remain calm. While it would be foolish not to keep watching the skies as 1950s sci-fi movies urge us to do, rationality demands that we now stop worrying unduly about them and learn to see them in a wholly new light. Meteorites pose little threat to us. Only one mass extinction event in all Earth history has ever been linked to one—and even then, one can be pretty certain that it was only part of a very complex story. We will never know for certain. But dinosaurs might well still be extinct today, even if the end-Cretaceous impactor were still aloft up there, quietly looping the Sun, minding its own business. We can say this because nothing significant in Earth history ever had a single cause.

Meteorites and their consequences, like all alien visitations, are mainly fashioned by our outlook on life, from the shadows of our own fears and from all the events and people we have known, which makes us what we are. I do not find it at all surprising that a man like Luis Alvarez, who not only helped develop the atom bomb but witnessed firsthand the destruction of Hiroshima, would have been so haunted by the prospect of sudden death from the skies. Nor does it surprise me that as a physicist, he would favor single, simple explanations over complex ones, or that given his combative personality and acquaintance with Edward Teller he fought so hard for his theory's acceptance and perhaps in the process overstated it. The full Alvarez is not invalidated because it may not, in the final analysis, prove to have been the whole truth. No scientific tale ever is.

Even when the objects themselves are not disputed, their interpretation may remain so. There are many ways to understand Maurizio Cattelan's *Ninth Hour*. While scientific stories are supposedly more constrained by observation and experiment, all our myths of the deep past, no matter how grounded we claim them to be, are always and forever creations of the human imagination that grow in their turn from our own context. The things that influence all storytellers influence scientists too—the same temptations and dangers lurk there for everyone. The need to obey narrative rules fights against the incompleteness of evidence, just as the desire to believe overcomes the need to remain rational, calm, always doubtful, ever ready to relinquish falsified ideas with joy. That this does not happen routinely, to say the least, reminds us that science remains a creative human endeavor.

In history, context is everything. During the chilliest days of the Cold War, when the specter of an implacable and almost unknown foe stalked the West, aliens were almost universally unpleasant. By contrast, at the close of the hippy generation, Steven Spielberg's tale of *Close Encounters of the Third Kind* (1977) conjured up a race of liberal, benign, childlike creatures, for whom to know all was clearly to forgive all. Yet not long after that, the skies grew dark again as an implacable, indestructible *Alien* (1979) burst out of John Hurt's stomach and shot across the breakfast table of the mining vessel *Nostromo*. The reignited Cold War was followed by the political dualities of the 1990s, when a serially contagious alien *Species* (1995) came wrapped in the comely exterior of model-turned-actress Natasha Henstridge; while *Independence Day* (1996) mobilized every positive and negative racial stereotype in Hollywood's unequaled repertoire to symbolize the human family's need to forget its differences in the face of rapacious alien asset-strippers bent on sucking the planet dry. More recently, *Avatar* (2010) placed our own species in this unattractive role, repeating on other planets the mistakes we have made and are making on Earth.

For centuries we humans have transplanted into meteorites the geological aliens, the heart of our own times, as we searched them for signs of times to come. To the peasants toiling in the fields, who were, until Siena, their only witnesses, the thunder-stone spelled doom and death; such was their sum expectation of life. To triumphant kings and priest emperors, they came trailing clouds of glory—signs from above that they, and no others, were not only the anointed ones here on this staging-post to the eternal life, but that in the celestial line-up each would take his rightful place at the deity's right hand.

But meteorites, the starry messengers, really lie far beyond our platitudes of good and evil. They link us to the cosmic environment in which we coexist, on a planet we all share. If poet, professor, peasant, emperor, subject, Elagabalus, Maximilian, Topham, Shipley, Banks, Soldani, Grewcock, and everyone who has ever held one of these oldest things that it is possible to hold can agree on one thing, it might be this: Our Earth is connected to the heavens, and everything upon it has roots that stretch back to the source of time, just as we find ourselves bound to our evolutionary origins upon our only home.

Whatever meanings we choose to derive from them, meteorites speak to us of the chance events within deep time in which we all have a common origin; and of the universe which, one way or another, is destined to reclaim our every atom.

ACKNOWLEDGMENTS

Many people have helped me in writing this book.

First, there are those who inspired me and guided my research. I am particularly grateful to Dr. Joseph McCall, through his work for the Geological Society of London and its magazine *Geoscientist*, which I edit, and on whose Editorial Board he is the longest serving member. Joe re-ignited my early fascination with space rocks, partly through his regular writing for the magazine and also through his organization of a 1997 Fermor Meeting at Burlington House on *Meteorites: Flux with Time and Impact Effects*. This superb conference, for which I did the media relations, led to my first meeting Gene and Carolyn Shoemaker. Gene, always so outgoing and unflappable, was a particular inspiration, and his tragic death in a car accident in Australia only months after, in which Carolyn, too, was seriously injured, touched us all. I am glad that a small morsel of his ashes finally made the journey he was denied, and now rests somewhere in the lunar regolith.

That meeting also introduced me to a host of others, including Dr. Robert Hutchison, Professor Monica Grady, Dr. David Rothery, Dr. Angela Milner, and many more. All have (largely unwittingly) helped to fan the flames of my enthusiasm for meteorites and their effects on life and Earth history, and my interest in the way we think about them as geological agents. The published work of these authors—and especially of the late Bob Hutchison—has been of inestimable value to me.

I need to thank yet again my undergraduate teacher and mentor, Professor Derek Ager. As well as being a leading light in the neocatastrophist movement of the 1970s, which rehabilitated the rare event in geological history, he was a pioneer in the field of palaeoecology, the discipline in which I was later to do research myself, and of course wrote that book which—by an absolute fluke—changed my life. However as Derek's undergraduate student, it was a privilege to watch him battle ultra-gradualist Lyellians, who had dominated geology for a century, and to do it with a vigor and sometimes venom that found expression in elegant and amusing English. This is a coincidence that in academia is as rare and as life-changing as the experience of staring down at a meteorite that survived as a fossil alongside the extinct creatures into whose world it dropped over 400 million years ago.

For particular and specific assistance I would like to thank the following: Professor Walter Alvarez; Sir David Attenborough; Dr. Emily Baldwin; Dr. Phil Bland; Professor Hilary Downes; Dr. Jack Meadows; Dr. Daniel Milner; Ather Mirza; Professor Colin Pillinger; Dr. Sara Russell; Professor Birger Schmitz; Lynda Smart (*Leicester Mercury*); Professor Fabio Speranza; Professor Hugh Torrens; Dr. Amarendra Swarup. Special thanks go to those who gave permissions and who critically read parts or all of the manuscript to save me from my solecisms. Any that remain are, of course, all my own responsibility.

Professor Gerta Keller of Princeton University deserves special mention because it was her hotly disputed work on the Chicxulub impact that provided me with the chance to write about her struggle to derail what she is convinced is a scientific bandwagon, providing me with a valuable lesson in the way that science works—or, sometimes, fails to work, especially when someone dares to question the hottest idea in town. Readers of my previous book *Supercontinent* will know that I like to explore relationships between science and mythology. Not only do meteorites provide ample opportunity for this, but they provide examples of how science as a culture creates its own myths: all of which, like many other myths, lie grounded in reality but which forever remain constructs of the human imagination.

No author can write a book without the help of an editor and a librarian. Of librarians I once again had the very best— Wendy Cawthorne, senior assistant librarian at the Geological Society of London. Wendy probably has her name in more books in that great library than almost anyone else apart from Darwin and Lyell, and the reason—unlike the material she manages to dig out—is not hard to find. Bella Lacey of Granta provided sound advice and guidance when the book was written and helped greatly to shape its final form, as did my agent, Peter Tallack.

Finally I should like to thank those closest to me for their help and encouragement, now and over the years—particularly my wife, Fabienne, for being my most exacting critic, and for acquiescing in more meteorite-related holidays than any husband has a right to expect.

FURTHER READING

There are scores of books about meteorites and mass extinctions—and many thousands of scientific papers and hundreds of research reviews. However, the following list comprises books that have been written with the general public or undergraduate student in mind, and which, with some indicated exceptions, present relatively few technical difficulties. I have listed only those which, in my background research for this book, have stood out for me—by virtue of their combination, in varying proportions, of breadth, style, readability, and general quality. It does not pretend to be a complete bibliography of this vast and inexhaustibly fascinating subject, and I offer my apologies to the authors of the many other excellent and useful books whose titles I have not had space to include here.

Many of the works listed are now out of print. However, I make no apology for listing them, even above those titles still to be found in bookshops, particularly since one of the greatest and most unexpected benefits of the new technology has been the ease with which one can now indulge in that old-fashioned sport of searching out and buying second-hand books.

METEORITE AND SPACE SCIENCE

Cokinos, Christopher, *The Fallen Sky: An Intimate History of Shooting Stars* (Tarcher/Penguin, 2009; ISBN: 978-1585427208). This attractive and readable book by a creative-writing teacher at Utah State University tells the story of (mostly) U.S. pioneers of meteoritics including Harvey Nininger, who in the eyes of the academic establishment never quite lived down the stigma of being a commercial collector. The author also visits Nördlingen in Germany, a town built within an impact crater, and participates in a meteorite hunting expedition to Antarctica.

Gribbin, John, *Stardust* (Penguin, 2000; ISBN 0-14-028378-1). If you want to learn more about stellar processes, their place in the evolution of the universe, stellar nucleosynthesis, and the way in which Fred Hoyle's inspiration led to the publication of the great Burbridge, Burbridge, Fowler, and Hoyle paper, this excellent and readable book by the acknowledged master of the popular science book in English is for you.

Hutchinson, Robert, *Meteorites: A Petrologic, Chemical and Isotopic Synthesis* (Cambridge Planetary Science, 2004; ISBN 978-0-521-47010-2). Bob Hutchinson's book sets the standard for comprehensiveness, covering theories about the early Sun, the protoplanetary nebula, the formation of the planets and a complete treatment of all known meteorite types. Extensive glossary and bibliography are included. Essential reading for any serious student of meteorites and their place in planetary science, but assumes an undergraduate-level grasp of a range of sciences.

Norton, O. Richard, *Rocks from Space* (Mountain Press Publishing Co., 2nd edn, 1998; ISBN 0-87842-373-7). Supremely readable and somewhat folksy romp through the whys and wherefores of meteorites and meteorite hunting. Halfway between a textbook and the more modest how-to type guide for the amateur meteorite enthusiast.

Wasson, John T., *Meteorites: Their Record of Early Solar-System History* (W. H. Freeman & Co., 1985; ISBN 0-7167-1700-X). Elegant textbook treatment unrivaled for its clarity and focus, with frequent summaries to keep the reader on track. Despite being a textbook, it appears to be telling a story; the only minor drawback being that the story is now somewhat superseded.

HISTORY

Burke, John G., *Cosmic Debris: Meteorites in History* (University of California Press, 1986; ISBN 0-520-05651-5). Masterly survey of the history of meteoritics from Aristotle to the late twentieth century, including chapters on curators and collections as well as folklore and myth associated with falls and fireballs. Monumental in scale and achievement.

McCall, G. J. H., Bowden, A. J., and Howarth, R. J., *The History of Meteoritics and Key Meteorite Collections: Fireballs, Falls and Finds* (Geological Society of London Special Publication No. 256, 2006; ISBN 978-1-86239-194-9). A collection of fascinating papers resulting from a conference at the Geological Society of London, covering a wide range of topics in the history of meteoritics and their curation. Notable for including a major overview of meteoritics in history by Ursula Marvin, doyenne of science historians.

EXTINCTIONS

Alvarez, Walter, *T Rex and the Crater of Doom* (Princeton University Press, 1997; ISBN 0-691-01630-5). Walt Alvarez's popular "how dad and I and all our friends did it" book. Highly readable personal account of the discovery of the iridium layer and the revelations that followed, and a textbook example of how to establish a new secular Bible story for the deep past.

Courtillot, Vincent, *Evolutionary Catastrophes* (Cambridge University Press, 1999; ISBN 0-521-58392-6). Vincent Courtillot is an inspiring scientist. This slim, riveting volume makes the case better than any for the linkage between major mass extinctions of the past with the eruption of Large Igneous Provinces, including the Deccan Traps of India, which culminated at the K-T Boundary. Courtillot also finds room to question theories of extinction periodicity. The English edition is updated from the original French one (Éditions Fayard, 1995) and contains relatively few scars of translation.

Glen, William, *Mass Extinction Debates: How Science Works in a Crisis* (Stanford University Press, 1994; ISBN 0-8047-2285-4). Collection of insightful essays, interviews, and transcripts by the foremost chronicler of a seminal period in Earth sciences.

Raup, David M., *The Nemesis Affair: A Story of the Death of Dinosaurs and the Ways of Science* (W. W. Norton & Co., 1986; ISBN 0-393-02342-7). University of Chicago paleontologist Dave Raup may never, he claims, have described a fossil species; but neither has he ever written a bad book. This account of his personal brush with the Nemesis hypothesis is the perfect insider view: wry, detached, intelligent—a sheer delight.

SIMPLE GUIDES

Dodd, Robert T., *Thunderstones and Shooting Stars: The Meaning of Meteorites* (Harvard University Press, 1986; ISBN 0-674-89137-6). This handy little volume is more book than guide but admirably summarizes the basics about meteorite science as it stood in 1986—which at this level is probably up-to-date enough for most readers.

Reynolds, Mike D., *Falling Stars: A Guide to Meteors and Meteorites* (Stackpole Books, 2001; ISBN 978-0-8117-2755-6). Friendly and enthusiastic how-to book for the amateur meteorite hunter and observer of meteor showers. Contains lists of contact organizations, museums, and dealers, almost exclusively based in the United States, as well as a timetable of meteor showers. Also covers basics of meteorite classes and tektites.

Russell, Sarah, and Grady, Monica, *Meteorites* (Natural History Museum, 2nd edn, 2002; ISBN 0-565-09155-7). Admirably concise and clear booklet in the NHM's wonderful series of short color guides, by two leading experts both of whom have held research positions at the Museum, which houses one of the world's greatest collections of meteorites.

Zanda, Brigitte, and Rotaru, Monica, *Meteorites: Their Impact on Science and History* (Cambridge University Press, 2001; ISBN 0-521-79940-6). Another short book inspired by a museum—this time by an exhibition at the Muséum d'Histoire Naturelle, Paris. This edition, consisting of separately authored chapters, is translated from the French original (1996). Covers history and science, impact craters and meteorites, and the origin of the Solar System.

RELATED INTEREST

Olson, Roberta J. M., and Pasachoff, Jay M., *Fire in the Sky: Comets and Meteors, the Decisive Centuries, in British Art and Science* (Cambridge University Press, 1998; ISBN 0-521-63060-6). Wonderful collaborative work, co-written by an astronomer and an art historian. Lavishly illustrated, detailing the scientific and artistic influence of comets and meteors from the dawn of the eighteenth century to the present day.

LIST OF ILLUSTRATIONS

Where no illustration credit is given, the illustration is in the public domain. The author and the publishers have made every effort to trace the copyright holders. Please contact the publishers if you are aware of any omissions.

ILLUSTRATIONS IN TEXT

p.9 Asteroid Belt. (Wikimedia Commons. Released to the public domain for use by any purpose without condition by Mdf at en.wikipedia.)

p. 17 Albrecht Dürer's *Melancholia*. (Photographic reproduction of original engraving, from Wikimedia Commons.)

p. 44 Earth's structure. (From an original in Lawrence Livermore National Laboratory Science & Technology Review, December 2007.)

p. 48 Major planets of the Solar System.

p. 50 Transverse section of Asteroid Belt. (Diagram by Piotr Deuar, released to Wikimedia Commons under the GNU Free Documentation License.)

p. 59 Reflectance spectra of asteroids. (Redrawn from an original in Norton, 1998: see Further Reading.)

p. 166 Derek Ager's "silly diagram." (Reproduced by permission of the Geologists' Association.)

p. 241 Diversity of species of brachiopod fossils through part of the Lower and Middle Ordovician in Baltoscandia. (Courtesy Prof. Birger Schmitz, redrawn from an original and reproduced by permission from Macmillan Publishers Ltd: Nature Geoscience Vol. 1 No. 1 © 2008.)

ILLUSTRATIONS IN INSERT

"Death Mask of Agammenon." (Photograph by Ted Nield.)

Professor Derek Ager. (Courtesy of Renée Ager, reprinted by permission.)

Luis and Walter Alvarez. (Courtesy Lawrence Berkeley National Laboratory.)

Barringer Meteor Crater, Arizona, USA. (Courtesy of John Underhill, reprinted by permission.)

"Figured stones" from Beringer's *Lithographiae Wirceburgensis*. (Reproduced in *The Lying Stones of Dr. Beringer*, 1963, University of California Press.)

Jean-Baptiste Biot (Lithograph by Auguste Lemoine. From www.sil.si.edu/digitalcollections/hst/scientific-identity/CF≠by_name_display_results.cfm?scientist≠Biot,%20Jean-Baptiste, Smithsonian Institution Libraries Digital Collection Scientific Identity—portraits from the Dibner Library of the History of Science & Technology.)

Ernst Chladni. (Reproduced courtesy of the Deutschen Staatsbibliothek, Berlin.)

Antoine de Lavoisier with his bride. (Portrait of Monsieur de Lavoisier and his wife, chemist Marie-Anne Paulze by Jacques-Louis David. Metropolitan Museum of Art, New York. From Wikimedia Commons.)

Varius Avitus Bassianus. (Photograph by Giovanni Dall'Orto of bust in marble in the Musei Capitolini, Rome.)

Grove Gilbert. (Courtesy of the Geological Society of London.)

The Ensisheim meteorite. (Image © Peter Marmet, www
.marmetmeteorites.com. Reproduced by permission.)

Imperial Hotel, Russell Square, Bloomsbury, London. (Non-
copyright image from www.imageshack.com.)

Boy struck by meteorite in 1992 in Mbale, Uganda. (Image by
unknown author from Dutch Meteorite Society, www.xs4all.
nl/~dmsweb/meteorites/mbale/mbale.html.)

Digby McLaren. (Courtesy of the Geological Society of London.)

Ernst Öpik. (Photo from collections of the Armagh Observatory,
http://star.arm.ac.uk/, released into the public domain
under GFD License.)

Lembit Öpik. (Photo by Salim Fadhley, 2006, released to
Wikimedia Commons under the Creative Commons
Attribution-Share Alike 2.0 Generic license.)

La Nona Ora 1999 by Maurizio Cattelan. (Photo courtesy,
Kunsthalle, Bonn © Maurizio Cattelan.)

Abbé Ambrogio Soldani. (Photograph by Ted Nield of original
portrait, Accademia dei Fisiocritici, Siena, Italy.)

Henry Sorby. (Reproduced by permission of the University of
Sheffield.)

Major Edward Topham. (Stipple Engraving by Peltro Tomkins,
published by J. F. Tomkins, after John Russell, February 28,
1790.)

Wold Cottage meteorite monument. (Photograph by Ted Nield.)

Hoba meteorite, Namibia. (Photograph by Ted Nield.)

INDEX

3C48 quasar galaxy, 29–30, 35, 43
132 Arethra asteroid, 53
433 Eros asteroid, 53
4600 Meadows asteroid, 3

Académie Royale des Sciences, 84
Accademia dei Fisiocritici, 75,
    130–31
accretion disks, 36–40
Aegospotami, 66
Agamemnon, 169–71
Ager, Derek, 95–99, 101–7, 121, 149,
    165–66, 205, 226, 242, 246, 260
    and brachiopods, 228–30, 238–39
AH84001 meteorite, 197–201
airliners, 10
Alexander VI, Pope, 22
algae, 237, 254–55
Alien, 265
Allan Hills, 197
Alsace, 18
aluminium, 39–40, 44
Alvarez, Luis, 123, 130, 134–35, 137,
    153, 163–64
    and K-T Boundary hypothesis,
    123–25, 130, 138–39, 152, 167,
    172–73, 189, 191, 195, 208–10,
    216–17, 231, 262, 264
    and nuclear program, 143–46,
    148, 171, 264

and Robert Oppenheimer, 155–56,
    167
Alvarez, Walter, 123, 130, 132–35,
    137, 149, 162
    and K-T Boundary hypoth-
    esis, 123–24, 130, 138–39, 152,
    172–73, 189, 191, 195, 208–10,
    216–17, 231, 262, 264
American Association for the
    Advancement of Science
    (AAAS), 109, 111
amino acids, 185, 187–88
ammonia, 45, 187
ammonites, 138, 173, 219
Anas platyrhynchos, 227
Anaxagoras, 21
Anders, Edward, 213, 216
Anders, William, 107
Andromeda constellation, 29–30
ANSMET Project, 198
Antarctica, 188, 197–98
Apennine mountains, 132
aphelion, 13
archaeocyathids, 254
archaeology, 169–70
Archaeopteryx, 192–94
Archimedes, 217
Aries constellation, 29
Aristotle, 65–66, 82
Armageddon, 154

Asaro, Frank, 123, 137, 139, 208
Astarte, 63
Asteroid Belt, 8–9, 13, 38, 47–52, 53, 58, 109, 214, 219, 259
   Kirkwood Gaps, 50–52, 54, 214–15
   position and appearance, 47–51
asteroid families
   Amor, 53–54
   Apollo, 53–54
   Aten, 53–54
   Baptistina, 259
   Flora, 214–15
   Gefion, 50, 215–16, 218–19, 233, 237, 239, 255
   Hirayama, 52
asteroid years, 49
asteroids, 8, 37–38, 40, 42, 46–54, 58–59, 109, 172
   discoveries, 46, 49
   distribution, 52
   groups, 52–53
   mass, 48, 109
   orbits, 49–51
   "reddening" of, 58
   "rubble piles," 52
astronomical units, 13
atomic energy, 141–43
atomic fission, 143–45
Attenborough, Sir David, 196
Australia, 152, 211, 234
Avalonia, 234, 243–44
Avatar, 266

B²FH paper, 34–35, 217
Bachelay, Abbé Charles, 84, 86
bacteria, magnetotactic, 99–201
Bakker, Bob, 125
ballistic science, 115
Baltica, 234–35, 243–44
Baltoscandia, 213, 240–41

Banks, Sir Joseph, 74, 76, 78–79, 266
barite, 207
barnacles, 223
Barringer, Daniel, 111, 114–21
Barringer, D. M., Jr., 119
Barringer Meteor Crater, 108–21, 165, 196
Barwell meteorite, 4–8, 12, 20, 26, 71, 74, 82
Basel, 16
Bathurst, Robin, Dr., 205
Bavaria, 186, 194
BBC television, 32
Beagle2 mission, 72
Benares, 79
Beringer, Johann, 186–87, 202, 219–20
Berthelot, Pierre, 181–82
beryllium-10, 135
Big Bang, 30, 33, 245
biodiversity, 201, 230–44, 248–49, 252–53, 255–57
Biot, Edouard, 87, 89
Biot, Jean-Baptiste, 87–89
biotite, 87
birds, 195, 227, 228
Birks, John, 146
Bland, Phil, 211
blastoids, 237
blood, 58
Bloomsbury, 140, 143
Boaden, James, 71
Bode, Johann Elert, 46–47, 49
Bonderoy, Fougeroux de, 84
Bonner, Thomas, 72
Born, Ignaz von, 79
Bottaccione Gorge, 130, 132–33
Bournon, Comte Louis de, 80
brachiopods, 229, 236–42, 256
bracken, 250–52
Brant, Sebastian, 20–21

Brazil, 161
Brazos River sections, 159–62
breccia, 44
    see also megabreccias
Bridlington, 71
Briggs, Raymond, 148
British Association for the
    Advancement of Science, 57,
    141, 183
Bronowski, Jacob, 156
Brunflo meteorite, 202–8, 211, 216
bryozoans, 237
Bryson, Bill, 104
Bunsen, Robert, 54–55, 57
Burbridge, Geoffrey, 35
Burbridge, Margaret, 35

calcite, 206–7
calcium, 39–40
calcium-aluminium inclusions
    (CAIs), 39–42
Cambrian Explosion, 232
Cañon Diablo Iron, 109–10, 117–18
Caracalla, Emperor, 62–63
Caravaca de la Cruz, 138
carbon, 31, 34–35, 40, 45, 109
    and extraterrestrial compounds,
        179, 181–82, 184–85
carbon dioxide, 36, 235, 237, 260
carbonates, 40, 78, 179, 199–200
Carroll, Lewis, 208
catastrophes, 96, 99, 103–4
    see also neocatastrophism
Cattelan, Maurizio, 22–24, 265
Cenozoic era, 134
cephalopods, 236
Chao, Edward Ching-Te, 119–20
Chaptal, Jean-Antoine, 87
Charig, Dr. Alan, 192
Chicxulub Crater, 154–55, 157–59,
    161, 163–64, 167–68, 244

China, 214, 218, 234, 243
Chinese astronomers, 16, 89
Chladni, Ernst, 19, 81–83, 86, 88
chondrules, 41–42, 54, 179, 206
Christianity, 15
chromite, 207, 212–13, 218
chromium, 207
Cilz, Marlin, 14
clams, 236, 238
Claringbull, Dr. G. F., 7
Clarke, Arthur C., 172, 262
Claus, George, 185
Clemenceau, Georges, 180–81, 183
climax communities, 248, 252
Clinton, Bill, 197, 199–200
Clöez, François, 181
Close Encounters of the Third Kind, 265
coal, 181
cobaltite, 207
Coes, Loring, 119
coesite, 119
Cold War, 147–48, 152–53, 156, 265
Collignon, Charles, 68
Collins, Gareth, 118
Colmar, 19
Colosseum, 63
Columbus, Christopher, 14, 18
comets, 16–17, 36, 39, 127, 139,
    150–53, 179, 182–83, 195
Compsognathus, 193
Connell, J. H., 253
conodonts, 236
conservation of matter, 85
continental drift, 98, 121, 125, 133,
    153
Copenhagen, 137–38
corals, 222, 254–55
cosmic microwave background, 33
cosmic radiation, 135
cosmic ray exposure (CRE) ages,
    213, 215

cosmic sediment, 42
Cosmochimica, 209
Cottonmouth Creek, 159
creationists, 168, 192, 196
Crick, Francis, 34
crinoids, 237
crocodiles, 8
Crow, Arthur, 6
Crutzen, Paul, 146, 148

Daniken, Erich von, 106
Darwin, Charles, ix, 30, 60, 167, 183, 192–94
Darwin, Erasmus, 29, 60
Day After, The, 148
De Laubenfels, Max, 106
de Luc, Guillaume-François, 72
Deccan Traps, 157–58, 163–64, 166, 260
deep sky, 29
denialism, 168
Derr, J., 12
Detroit, 263
diamonds, microscopic, 109
Dickens, Charles, 69
dinosaur skeletons, 206
dinosaurs, extinction of, 5, 122–26, 134, 137–39, 154, 158–59, 189, 216, 219, 258–59
Dixon, Adele, 32
DNA, 34, 195
Dr. Strangelove, 145
Dournon, battle of, 18
dropstones, 204
Dürer, Albrecht, 16–17
Dyrenforth, James, 32

Earth, 40, 44–46
    accretion, 45
    age, 164, 168, 183
    chemical composition, 136
    early atmosphere, 187
    geological activity, 44, 186
    magnetic reversals, 128, 133
    mass, 45
    nickel-iron core, 44, 80, 136
    rotation, 38
    structure, 44
earthquakes, 263
Earthrise picture, 107
Ecclesiastes, 97
ecological science, 246–53
Eddington, Sir Arthur, 257
Einstein, Albert, 34, 145, 190
Elagabalus, Emperor, 62–64, 89, 252, 266
electromagnetic radiation, 32
electrophonic sound, 12
Emesa, 61–62, 64
Empire Strikes Back, The, 47
Enclosure Acts, 90, 250
Engelder, Terry, 134
England, Percy, 6–7, 14
Ensisheim Thunderstone, 14–22, 26
Eocene Period, 8
Estonia, 150–51
European Space Agency, 72
evaporites, 235
evolution, 164, 167–68, 177, 246
    and Hoyle's Fallacy, 192–97
    Lord Kelvin and, 182–84
Explorer 1 and 2 satellites, 128
extinction points, 5, 11–12
Extra-Terrestrial Exposure Law, 177, 185

Fennoscandian Shield, 203
Fifield, Richard, 150
Flammarion, Camille, 192
Flammarion, Ernest, 192
floods, 105
flowering plants, 126

Foote, Arthur, 109–11
foraminifera, 131, 133–34, 162
Fowler, William, 35
French Revolution, 68, 79, 83, 85–86
Freud, Sigmund, 166
Friedrich III, Holy Roman Emperor, 18
Frisch, Otto, 143

Gaia, 173
Galileo, 168
Ganges River, 79
gardening, 44
Gassicourt, Cadet de, 84
Gentleman's Magazine, 71
Geochimica et Cosmochimica Acta, 209, 221
Geological Society of America, 121
Geological Society of London, 80, 103, 162, 169
Geology, 209, 221
George, King, of Greece, 170
George III, King, 69
Gibbon, Edward, 63
Gibbons, Jack, 197
Gibson, Mel, 13
Gilbert, Grove, 111–16, 119–21, 165, 196
Gillray, James, 69
glaciers, 204
glass harmonica, 81
Glen, William, 171
global warming, 126–27, 168
Globotruncana, 132, 134
Gondwanaland, 234–35, 237, 242
Gorbachev, Mikhail, 147
Gordon Riots, 69
Gotland, 95–97, 166, 221, 253
Gould, Stephen Jay, 104, 140, 171, 226
Gower Peninsula, 247

Granby crater, 213
Grand Tour, 68, 74–75
graptolites, 235–36
Graves, Robert, 156
Great Ordovician Biodiversification Event (GOBE), 232–44, 255–57
greenhouse climate, 235, 242
Greenland, 24
Greg, Robert, 57–58
Gregory, Graham, 31
Gregory, Olinthus, 87
Greville, Charles, 74, 79
Grewcock, Joseph, 6, 266
Gubbio, 130, 132, 137–38
Gulf of Mexico, 154
Gutenberg, Johannes, 20
Guy, Birdie, 25

Haber Process, 142
Häberlein, Dr. Karl, 192
Haldane, J. B. S., 257
Hale-Bopp comet, 17
Halley, Edmond, 127
Halley's comet, 17
Hamilton, Emma, 74, 79
Hamilton, Sir William, 73–76, 78
Hamilton, Vicky, 198
Hardy, Thomas, 8, 24
helium, 31, 33–35, 38, 45, 145, 245
Helmholz, Hermann von, 182
Hennessy, Douglas, 185
Henry, Thomas, 143
Henstridge, Natasha, 265
Herculaneum, 74
Herodian, 62
Herschel, William, 46, 78, 88
Hervey, Frederick, 74–76
Heston, Charlton, 154
Hierocles, 64
high explosives, 142

Higham, Norman, 56
Hind, John, 37
hippopotamuses, 8
Hirayama, Kiyotsugu, 52
Hiroshima, 144, 264
Hitler, Adolf, 140
HIV and AIDS, 168
Hodges, Ann and Hewlett, 25
Hodgkinson, Richard, 107
Högklint Limestone Formation, 254
Holy Roman Empire, 18, 20
Homer, 169–70
Homs, 61
Hooke, Robert, 66
Howard, Edward, 79–80
Hoyle, Sir Fred, 35, 189, 191–97, 217
Hoyt, William, 120
hurricanes, 103, 105
Hurt, John, 265
Hutton, James, 100–102
Huxley, Thomas Henry, 103, 184, 193, 257
Hyades Cluster, 37
hydrocarbons, 181, 185, 199
hydrochloric acid, 260
hydrogen, 31, 33–35, 37–38, 45, 85, 145, 181, 187, 245

Iapetus Ocean, 205, 214, 244
ice ages, 237
icehouse climate, 237, 242
impact metamorphism, 120
Independence Day, 265
influenza epidemics, 195
Innisfree meteorite, 53
Innocent VIII, Pope, 22
Institut National, 87
intelligent design, 183
intermediate disturbance theory, 253
iridium, 43, 135–39

iridium anomaly, 137–39, 145–46, 157, 159–60, 166, 209, 262
iron, 31, 34–35, 40, 43–44, 57, 78, 86, 109, 111–12, 132, 136
    see also nickel-iron
iron sulfide, 40, 77, 86, 207
Ivuna meteorite, 185

Jämtland, 205
John Paul II, Pope, 22
Johnson, Willard, 111
Julia Domna, 62
Julia Aquila Severa, 64
Juncus conglomeratus, 186
Jupiter, 8–9, 38, 45–52, 259
Jurassic Park, 152

Kaaba, 64
kangaroos, 152, 256
Kärdla crater, 213
Kazakhstania, 234
Keith, Patsy, 12
Keller, Gerta, 157–65, 168–69, 244
Kelvin, William Thomson, Lord, 182–84
King, Edward, 73, 76, 78
Kirchoff, Gustav, 54, 57
Kirkwood, Daniel, 51
Knapp, Michelle, 8, 13–15, 24, 51, 54
Krahut meteorite, 79
K-T Boundary, xiv, 134–36, 138, 152, 157, 159–61, 195, 208–10, 216, 231
Kubrick, Stanley, 145
Kuiper Belt, 36

L'Aigle meteorite, 87–88
Lang, Allan, 14
Laplace, Pierre-Simon, 127
Large Igneous Provinces, 157, 260
Laurentia, 234, 236, 243–44

Lavoisier, Antoine-Laurent de, 84–86
Le Grand Lucé meteorite, 84
Le Mans, 84
Leicester Mercury, 7
Leipzig, 81
Life magazine, 25
lignite, 181
limestones, 193–94, 202–5, 207–8, 210, 212, 218
Lipmann, Charles, 185
lithium, 33
Lloyd Williams, John, 79
Lockne crater, 213
London Chronicle, 73
Lost City meteorite, 53
Lowrie, Bill, 132–33
Lucignano d'Asso, 131
Lucretius, 88, 150
Luther, Martin, 104
Lyell, Charles, 101–3, 134

Ma Tuan-Lin, 89
McKay, David, 197, 199–201
McLaren, Digby, 106, 127–29
Macrinus, Emperor, 63
magnesium sulfate, 179
magnetite, 179, 199–201
mammals, 8, 195
Manhattan Project, 143, 145
Mars, 8–9, 26, 40, 45–49, 72
    Eos Chasma, 198
    and evidence for life, 197–201
    iron core, 40
    Isidis Planitia region, 72
    magnetization, 201
    orbit, 53
    Valles Marineris system, 198
Mars Express, 72
Marvin, Ursula, 16
Marx, Karl, 104

mass extinctions, 99, 105–6, 123–24, 139, 152, 165–67, 177, 231, 258
    and comets, 152
    cyclicity of, 149, 153, 231
    end-Cretaceous (K-T), xi, xiv, 96, 122–39, 152–67, 195, 231, 240–41, 244, 249, 259–64
    end-Ordovician, 237
    end-Permian, 127–28, 129, 233
    and nuclear winter theory, 145–48
    and recovery phases, 240
Matthews, Drummond, 129
Maupertuis, Pierre-Louis de, 127
Maximilian I of Hapsburg, 18–21, 266
Mbale, Uganda, 24, 214
Meadows, Jack, 3–6
Mecca, 64
Medawar, Sir Peter, 120, 122
megabreccias, 243–44
Meinschein, Warren, 185
Meitner, Lise, 143
Melosh, Jay, 118
Mercury, 8, 10, 40, 45–46
    The Spider structure, 110
Merrill, George, 116–17, 119
Mesozoic Era, 134, 166, 233
meteorite flux, 211–13, 216, 218, 240, 242, 244, 246
meteorite impact craters, 107–8, 150, 198
meteorites
    age, 7–8
    chemical composition, 39, 77–78, 80, 136
    formation, 40–45
    mineralogical separation, 77–80
    oriented, 62
    origins, 51–54, 58, 80–81
    temperatures, 86
    vaporization, 117

meteorites, types of
  carbonaceous chondrites, 45, 178, 182, 188
  CI, 179
  diogenites, 197
  L chondrites, 210–11, 213–16, 218–19, 233, 239–40, 242
  H chondrites, 211
  ordinary chondrites, 41
  Shergottite, Nakhlite and Chassigny (SNC), 197
Meteoritical Society, 40, 264
methane, 36, 38, 45, 187
Meyer, Nicholas, 148
Meyer, Ray, 14
Michel, Helen, 123, 137, 139, 208
microfossils, 130, 133, 159, 161
micrometeorites, 212–14
micropaleontology, 131
microscopy, 56–58, 202
mid-Atlantic Ridge, 128
Milky Way, 29, 35
Miller, Arnold, 226
Miller, Stanley Lloyd, 187
Milner, Dr. Angela, 192
Mittlefehldt, David, 197
Montauban, 178, 184, 186–87
Moon, 45, 78, 109
  craters, 66, 111
  mass, 109
Moon landings, 177
Moore, Raymond, 229
Mordaunt, Thomas, 103
Morrison, Philip, 143
Mount Eibelstadt, 219
Mount Vesuvius, 66, 75–76, 78, 102
Mozart, Wolfgang Amadeus, 79, 81
Mullinax-1 core, 159
Multiple Working Hypotheses, 112
mummification, 126
Mycenae, 170

Nagy, Bart, 185–86
Nakhla dog, 26, 263
nanobacteria, 199
Naples, 73, 76–77
Napoleon Bonaparte, 83
NASA, 107
  and Mars, 199–200
  Spaceguard program, 156, 172, 262–63
National Geographic, 120
Native Americans, 108
Natural History Museum, London, 7, 192, 211
natural selection, 167, 195–96
Nature, 208
nautiloids, 207, 214
Near Earth Objects (NEOs), 53–54, 172, 262
Nelson, Lord Horatio, 74
Nemesis star, 108, 150, 152–53, 165
neocatastrophism, 104, 107, 123, 158, 173, 205, 228
Neptune, 47
neutrons, 142
New Jersey, 161
New Scientist, 139, 150, 191–92
New York Times, 122, 184–85, 197
Newton, Sir Isaac, 54, 66, 82, 125
Nicholaston, 247
nickel, 40, 43, 80, 114, 136
nickel-iron, 44, 109, 114, 118, 178, 207
Nietzsche, Friedrich, 80
Nininger, Harvey, 118
nitrogen, 34, 142, 181–82
Nobel Prize winners, 124, 146, 148, 168, 187
northern hemisphere, 58
nuclear weapons and war, 143–48, 171

nuclear winter theory, 145–48, 153–54
Nullarbor Plain, 211

Occam's razor, 107, 114, 165, 193
Olbers, Heinrich, 82
olistostromes, 243
olivine, 40, 206
Oort, Han, 151
Oort Cloud, 36, 39, 108, 151–53
Öpik, Armin, 151
Öpik, Ernst, 150–52, 158, 172, 256
Öpik, Helgi, 151
Öpik, Lembit, 151, 172
Oppenheimer, J. Robert, 143–45, 155–56, 168
Ordovician period
  brachiopod diversification, 237–41
  continents, 233–35, 242, 256
  Great Ordovician
    Biodiversification Event
    (GOBE), 232–44, 255–57
  increased meteorite flux, 202–18, 221–24, 249, 257, 260
Orgueil meteorite, 177–79, 181–82, 184–87, 193
Orion Nebula, 37
orthocones, 210
osmium, 43, 135
Owen, Sir Richard, 193
Oxford University, 76–77
Oxwich River, 247
oxygen, 34–35, 85, 181–82, 187, 197, 235
ozone layer, 11, 146

Palaeocene–Eocene Thermal
  Maximum (PETM), 8
palaeoecology, 97–98, 224, 238, 274
palaeomagnetism, 234
Paleobiology, 223–24

Paleontological Society of America, 106, 127
panspermia, 180–81, 183–84, 189, 196
Papua New Guinea, 263
Parnell, John, 242–43
Pasteur, Louis, 180–83, 185–88, 218
Paulze, Marie-Anne, 84
Pausanias, 170
Peekskill meteorite, 9–13, 16, 20, 26, 51, 53
peer review, 164, 190–92, 223
Pennant Hills, 98, 249–50, 252
periodic table, 30–31, 33–35
Perseus constellation, 29
Petróleos Mexicanos (PEMEX), 154
phengite, 207
Piazzi, Giuseppe, 46, 49
Pidgeon, Walter, 129
Piiri, Alide, 151
Pillinger, Colin, 72–73
Pisces constellation, 29
Pius XII, Pope, 25
planetesimals, 43–44, 111
planets
  chemical composition, 57
  gas giants, 8, 36–38, 45
  orbits, 46–49
  origin, 26, 36, 40, 42, 127
  rocky, 8, 36–38, 40, 44–45, 51
  and Titius–Bode Law, 47
plate tectonics, 99, 121, 133, 135, 171–72, 245
Platts, Harold, 7
Pliny the Elder, 21, 66
Plutarch, 66
Pluto, 36, 48
plutonium, 137–38, 143–44
Poland, 23
polycyclic aromatic hydrocarbons
  (PAHs), 199
Pompeii, 74

Pouchet, Félix, 180–81
printing, 20
Puxi River, 214, 218
Pyramids, 135
pyroxene, 40, 207

quasars, 29

radiation, 128–29, 135
radio, 183
radioactive decay, 42–45
Raup, David, 149, 153, 226, 230–31
Reagan, Ronald, 147, 156
red shift, 30, 33
reflectance spectra, 55, 58–59, 214
regolith, 43–44
Reubens, Paul, 13
Revolutionary War, 69
rhinoceroses, 8
rhynchonellids, 238
Riga, 81
RMS *Titanic*, 24
Rome, 24
Roosevelt, Theodore, 114
Rottenberg, Anda, 23
Rowlandson, Thomas, 69
Royal Society, 58, 73–74, 76, 82, 127
rudists, 254
Rutherford, Ernest (Lord), 124,
    141–43, 169

Sagan, Carl, 146–47, 152
St. Francis of Assisi, 132
St. Paul, 217
St. Peter, 24
Sainte-Beuve, Charles, 87
Salisbury Crags, 100
San Francisco Chronicle, 143
San Salvador, 18
Sardinia, 238
Saturn, 38, 46, 49

Saussure, Horace-Benedict de, 56
Sawdon, George, 70
Scaglia Rossa, 133–34
Schilling, Diebold, 16
Schindewolf, Otto, 128, 137–38
Schliemann, Heinrich, 169–71
Schmitz, Birger, 208–16, 218–19, 221,
    223–24, 239, 242–43, 256–57
Science, 147, 194, 197, 253
scleractinians, 254
Scotland, 243
sea levels, 161, 235, 237, 254, 256
Sepkoski, Jack, 149, 153, 224–26,
    229–32
Shaw, Quincy, 117
Sheffield, 41, 54, 56–57, 148, 202
Shipley, John, 70–71, 91, 266
Shiva, 150
shock metamorphism, 109
shocked quartz, 210
Shoemaker, Eugene Merle, 119–220
Siberia, 234
siderophile elements, 43, 136
Siena meteorite, 73–79, 130–31, 258,
    266
Sigismund of Austria, Archduke, 18
silica, 115
silicates, 40, 43–44, 136
silicon, 31
Siljan Ring, 206
Silurian reefs, 221–22, 237, 254–55
skeletons, 222–23, 232
Smit, Jan, 138–39, 162–63, 169
snowline, 36, 38, 45
Socrates, 100
sodium, 55
Soergel, Philip, 19
solar composition, 36–37
solar spectrum, 37
Solar System
  age, 39

chemical composition, 57, 136
formation, 29–30, 34, 36, 42–43, 60
Late Heavy Bombardment, 199
regularity, 47
solar wind, 17, 37, 39, 41
Soldani, Abbé Ambrogio, 75–78,
    130–31, 133, 266
Solnhofen Limestone, 193
Sorby, Henry, 41, 54, 56–58, 202
southern hemisphere, 233
Soviet Union, 145
Space Shuttle, 10
space weathering, 58
species, 222–25, 227–29
Species, 265
spectrum analysis, 54–55, 57
spherules, 159–61
Spielberg, Steven, 265
sponges, 222, 237
spontaneous generation, 179–81,
    183, 186
Standard Iron Company, 114
Stardust Mission, 39
stellar nucleosynthesis, 31, 34–35,
    191, 217, 246
stereochemistry, 188
Sterne, Laurence, 68
Stevns Klint, 137
Stockholm, 95
Strasbourg, 20
Stromatactis, 205
stromatoporoids, 222–23, 237,
    254–55
subduction, 133, 233
sulfates, 40, 179
sulfides, 179
sulfur, 13, 15, 40, 71, 158
sulfur dioxide, 260
Sun
    chemical composition, 36–37, 57
    hydrogen conversion, 34

origin, 34–36, 127
rotation, 38
sunlight, 54, 57–58
supernovas, 34–36, 43, 128, 137–38
Swansea, 97–98, 249
syphilis, 22
Syria, 61–63
Szilárd, Leó, 140–44

Tanzania, 185
tartaric acid, 188
Tassinari, Mario, 210
Taurus constellation, 37
taxonomy, 227–29
tektites, 129
television, 32
Teller, Edward, 145, 147, 155, 264
Thames, River, 8
Thatcher, Margaret, 147
thermohaline circulation, 235
Thirty Years War, 19
Thomas, Dylan, 131
Thomson, Guglielmo, 76–78, 80, 130
Thorsberg Quarry meteorite, 208,
    210–11, 218
Thorslund, Per, 95–96, 166, 201–2,
    204–8, 211, 216
Thrace, 66
Threads, 148
threshold concepts, 216
Tiber, River, 61
Tilghman, Benjamin, 115, 119
Times, The, 72, 141
Titius, Johann, 46, 49
Tizio, Sigismondo, 22
Topham, Edward, 67–73, 75, 82,
    90–91, 266
Torrens, Hugh, 77
Treaty of Senlis, 18
T. rex, 126, 166, 260–61
Triangulum constellation, 29–30, 43

trilobites, 236
troilite, 207
Trollope, Anthony, 102
Trondheim, 244
Trower, W. Peter, 155
Troy, 169
tsunamis, 105, 159–61, 163, 243
TTAPS report, 147
T-Tauri stars, 37
Tucci, Stanley, 13
Tvären crater, 213
Twain, Mark, 79

ultramafic rocks, 202–3, 205
uniformitarianism, 15, 100–104, 123,
    197
United States Geological Survey,
    111, 116, 119, 120
Uppsala, 95
Urania, 63
uranium, 143–44
Uranus, 38, 46–47, 78
Urey, Harold, 129, 187
US science, 112

Van Allen, James, 128
Van Allen Belts, 128–29
vanadium, 207
variable stars, 37
Varius Avitus Bassianus, 62
Vatican, 22–23
Velikovsky, Immanuel, 106
Venus, 8, 10, 40, 45–46
vertebrate paleontologists, 123–25
Victoria asteroid, 37
Vine, Fred, 129
Visby Formation, 221–23, 254
Voltaire, 104
Vosges mountains, 15–16

Walker, Edmond, 14

Walley, Chris, 149
Washington Evening Star, 143
water, 36, 38, 179
water vapor, 235
Watson, James (astronomer), 53
Watson, James (groom), 70
Watson, James D., 34
Watts, Nigel, 222, 254
Wegener, Alfred, 153
Wells, H. G., 142–43, 169, 182, 195
Wendt, Herbert, 196
Whewell, William, 102–3
Wickman, Frans, 206
Wickramasinghe, Chandra, 192–96
wildfires, 250
Wild-II comet, 39
Willis, Bruce, 172
Wister, Owen, 114
Wold Cottage meteorite, 70–73, 76,
    79, 90
Wordsworth, William, 246
World War I, 142, 251
World War II, 142

X-rays, 183

York, 67–68
Yucatan Peninsula, 154, 161
Yucca Flats, Nevada, 119